HOMESTEADING MASTERY: ADVANCED SKILLS AND TECHNIQUES ON THE MODERN HOMESTEAD

BY

AUDRA HALDANE

HONEY SHACK

BOOKS IN THIS SERIES

The Homestead Blueprint: Introductory Self-Sufficiency With Homesteading

Homesteading Mastery: Advanced Skills and Techniques on the Modern Homestead

Homesteading for Total Self-Sufficiency

The Urban and Suburban Homestead

Homesteading and Natural Building Techniques

Sustainable Homestead Design and Permaculture Principles

Finding Your Homestead: A Comprehensive Guide to Purchasing Land in the U.S., Canada, and Europe

TABLE OF CONTENTS

INTRODUCTION

Are you ready to begin the next chapter of your homesteading journey?

You've laid the foundation—tending the garden, raising animals, preserving food, and embracing self-sufficiency. Now, you're ready for the next step: scaling up, refining your skills, and integrating advanced techniques that will elevate your homestead into a true powerhouse of productivity, sustainability, and personal renewal.

The truth is, mastering homesteading isn't just about doing the work—it's about doing it smartly and with intention. It's about bringing together the best of traditional knowledge and modern innovation to create a homestead that works for you. This book is designed specifically for you: the homesteader who already has a solid understanding of the basics and is now ready to tackle the challenges that come with going bigger, better, and more self-sufficient.

In the pages ahead, I'll guide you through the advanced techniques that can take your homestead to the next level. From adapting your practices to different climates and scaling your garden to developing sustainable energy systems, each chapter is a treasure chest of strategies, insights, and practical tips to solve the pain points you're facing right now. Whether you want to increase your honey production, maximize water efficiency, or learn how to make the most of the wild foods around you, this book is your go-to resource.

This is **not** a book for beginners. It's for the homesteader who's ready to work smarter, not harder, and who's seeking ways to enhance and expand the systems that are already working on their homestead. You've done the hard work, and now it's time to make your homestead more efficient, sustainable, and financially viable. Think of this book as your roadmap to becoming the homesteader you've always wanted to be—one who is truly self-sufficient, running a homestead that is as productive as it is peaceful.

Scaling to New Heights

Maybe you're already growing a successful garden, but you're itching to learn how to expand your growing space and make the most of your land. Or perhaps your animals are thriving, but you're looking for advanced strategies to take your husbandry to the next level—whether that's breeding for specific traits or raising more diverse livestock. You might even be ready to integrate renewable energy into your homestead to cut utility costs and lessen your dependence on the grid. No matter where you are in your homesteading journey, this book offers you the solutions and techniques you need to take things to the next level.

We'll start by addressing the challenges and opportunities presented by different climates. Understanding how your environment influences your crops, animals, and energy needs is crucial to scaling your homestead in a way that works with—not against—nature. We dive deep into homestead design and layout, showing you how to optimize your space for efficiency and productivity.

Once you've mastered the land and layout, we move on to discuss expanding your garden in creative and innovative ways. Growing more food doesn't have to mean working harder—I'll show you techniques to maximize yield with less effort.

From there, we tackle advanced animal husbandry, offering methods that will allow you to increase livestock health, production, and sustainability. For those who keep bees, I'll teach you how to maximize honey production and increase pollination efficiency, turning your apiary into a well-oiled, thriving system.

Energy independence is crucial for any modern homesteader, and I'll guide you through the world of renewable energy, offering practical solutions for harnessing solar, wind, and hydro power in ways that are both affordable and effective.

Food storage and preservation techniques are also covered. Preserving your harvest for the long term is an art and science in itself, and I'll provide you with advanced methods to ensure you have a stockpile of food that lasts through every season.

Then we'll delve into wildcrafting and foraging—learning how to tap into the wild food sources around you to supplement your homestead's food supply in a sustainable, safe, and responsible way.

Water is life on the homestead, and I'll teach you how to master water management systems, from rainwater harvesting to irrigation systems that conserve water and optimize its use on your property.

When it comes to infrastructure, we'll cover everything from building barns and greenhouses to building an outdoor still and other essential structures that will enhance your homestead's efficiency.

Your homestead isn't just a place to grow food—it's a sanctuary. We'll explore how to create spaces on your homestead that nourish both your body and mind, offering practical ideas for outdoor retreats, nature-inspired wellness spaces, and using herbs and plants for DIY beauty treatments.

We'll integrate technology with homesteading, discussing smart tools and systems that can help you manage your homestead more efficiently. From farm management software to solar-powered gadgets, technology has a place on the modern homestead.

We'll also explore how to generate income from your homestead, whether through selling produce, homestead products, or setting up your own homestead-based business. This chapter will show you how to leverage the skills and products from your homestead to create a sustainable income.

This book ends with **Chapter Fourteen**, which wraps up everything we've covered, helping you transition to a fully self-sufficient lifestyle where you're in charge of your food, energy, and income.

As you read through the chapters, you'll find that the strategies I provide are actionable, rooted in years of experience, and customized for homesteaders who are already familiar with the basics. The common pain points of expanding a homestead—lack of time, space, or resources—will be solved through practical advice that addresses the realities of homesteading in today's world.

Scaling your homestead doesn't have to be overwhelming or exhausting. It's about working smarter, not harder. The tips, tricks, and strategies in this book will empower you to maximize your resources, minimize waste, and create a homestead that is more productive, resilient, and deeply fulfilling.

By the end of this book, you'll not only have the technical skills and knowledge you need to expand your homestead, but you'll also have the confidence to embrace the future of homesteading—an era where technology, sustainability, and creativity come together to create a truly modern, self-sufficient lifestyle.

So, grab your favorite pen and notebook, get ready to take action, and let's dive into the exciting world of advanced homesteading together!

ABOUT THE AUTHOR

Audra Haldane is a lifestyle writer and passionate advocate for self-sufficiency whose expertise spans from the sprawling landscapes of Appalachia to the misty hills of Scotland, where she operates her own farms.

Known for her accessible, hands-on approach to homesteading, Audra's books—including *The Homestead Blueprint: Introductory Self-Sufficiency With Homesteading*, *Advanced Homesteading Skills and Techniques*, and *Sustainable Homestead Design and Permaculture Principles*—have become essential reads for those looking to cultivate independence, resilience, and sustainability.

When she's not tending to her farms or sharing practical homesteading wisdom, Audra indulges her love for travel and reading, constantly seeking inspiration from new places and stories.

With her guidance, readers from urban high-rises to remote countryside homes discover the joys of sustainable living, making her one of the most beloved voices in the modern homesteading movement.

CHAPTER ONE: DIFFERENT CLIMATES

Tailoring homesteading practices to specific climates is the difference between a thriving, resilient homestead and one that struggles to meet its potential. In this chapter, we'll explore strategies to adapt to diverse environmental conditions, tackle region-specific challenges, and turn climate from a challenge into an asset.

Growing Seasons—Nature's Non-Negotiable Rules

Understanding the growing seasons in your region is foundational for advanced homesteaders. You're not just working with the land—you're collaborating with it.

COLD NORTHERN REGIONS

In northern climates, the growing season might be a fleeting 3-5 months. Every day counts, and efficiency is key.

Season Extension Tools:

Hoop Houses: Constructed using PVC pipe or metal hoops, these structures create a microclimate that can add weeks to your growing season. For example, start hardy greens like spinach and Swiss chard under a hoop house in early spring while snow still lingers.

Row Covers: Lightweight fabrics that protect tender seedlings from frost and pests. They are especially effective for crops like brassicas.

Thermal Mass Heating: Place black barrels filled with water inside greenhouses. These absorb heat during the day and release it at night, helping to stabilize temperatures.

Crop Timing:

Invest in short-season cultivars. For instance, "Alaska" peas and "Early Girl" tomatoes mature faster than standard varieties.

Use succession planting. After harvesting radishes (which mature in 30 days), follow up with a second crop like lettuce or spinach.

ARID DESERTS

The desert is unforgiving but incredibly rewarding when approached strategically.

Water Harvesting:

Passive Rainwater Capture: Build swales (shallow trenches) along the contour of your land to slow and capture runoff. These can recharge underground water and irrigate trees planted alongside.

Greywater Systems: Divert water from sinks, showers, and washing machines (with non-toxic soaps) to irrigate non-edible plants like fruit trees.

Mulching: Cover soil with organic materials like straw or wood chips to reduce evaporation by up to 75%.

Plant Strategies:

Use desert-native species like mesquite or agave for windbreaks and erosion control. Mesquite provides pods that can be ground into a protein-rich flour.

Practice interplanting with nitrogen-fixing plants like desert senna to improve soil fertility.

TROPICAL AREAS

Tropical climates come with generous rains and heat but also relentless pests and diseases.

Crop Rotation: Rotating crops like sweet potatoes, taro, and ginger can disrupt pest cycles and improve soil health.

Shade Management: Plant multi-story layers, such as taller bananas and papayas to create shade for understory crops like lettuce or beans. This mimics a natural rainforest canopy.

Disease Control: Use organic fungicides like copper sulfate to combat mildew and rot, which are common in humid environments.

The Right Plants and Animals for Your Climate

Selecting region-specific crops and livestock ensures efficiency and sustainability. This section dives into advanced strategies for choosing and managing homestead inhabitants.

COLD NORTHERN REGIONS

Crops: Grow perennials like horseradish and sunchokes that come back stronger each year. Pair them with hardy annuals like beets and kale for diversity. Incorporate "C3" plants.

Livestock: Opt for breeds like Highland cattle, which have thick coats and thrive in cold. For chickens, consider breeds like Buff Orpingtons, known for their cold-weather laying.

Pro Tip: Build deep-litter bedding systems in animal coops. Layers of straw and manure compost in place, generating heat for animals during winter.

ARID DESERTS

Crops: Prioritize heat-tolerant varieties. Cowpeas and millet are excellent staples that require minimal water. Incorporate "C4" plants.

Livestock: Choose drought-tolerant animals like Barbary sheep or dromedary camels. Camels, for instance, can produce milk and meat while surviving on sparse vegetation.

Pro Tip: Plant fodder crops like alfalfa or cactus pads as supplementary feed.

TROPICAL AREAS

Crops: Incorporate "C4" plants that grow quickly in high light and heat, like cassava and sugarcane. Pair with legumes like pigeon peas to fix nitrogen.

Livestock: Ducks thrive in wet conditions and help control insects in rice paddies or garden beds.

Pro Tip: Create a permaculture food forest with layers of productive plants—trees for fruit, shrubs for berries, groundcovers for mulch, and vines like passionfruit for vertical space.

Climate-Specific Challenges

Advanced homesteading means thinking ahead and solving problems before they arise.

COLD NORTHERN REGIONS

Greenhouse Design: Consider geothermal greenhouses that tap into the earth's consistent temperature to grow food year-round.

Energy Tips: Insulate barns with straw bales and use reflective panels to direct sunlight into animal housing during winter.

Arid Deserts: Beating Heat and Erosion

Soil Building: Implement hugelkultur beds, which use buried wood to retain moisture and improve soil structure over time.

Wind Control: Plant fast-growing trees like Moringa along the periphery of your property to create living windbreaks.

TROPICAL AREAS

Water Management: Install French drains around your home and garden beds to prevent flooding.

Pest Strategies: Encourage beneficial predators

like frogs by building small ponds or wetlands. These natural allies will keep insects under control.

C3 Plants for Northern Cold Climates

In northern, cold climates where winters can be long and summers short, C3 plants are your best friends for food production. These plants thrive in cooler temperatures, making them perfect for areas where the growing season is limited. Understanding how C3 plants work, how to choose them, and how to manage them will help you make the most of your short summers and cold growing conditions.

WHAT ARE C3 PLANTS?

C3 plants are the most common type of plants on Earth. They utilize the C3 photosynthetic pathway, which is the most efficient in cooler temperatures and moderate light conditions. C3 plants do best in temperate climates where the temperatures are cooler (usually between 55°F and 75°F / 13°C and 24°C) and water is relatively abundant. These plants perform photosynthesis at a moderate rate, making them ideal for cold northern regions where extreme heat and drought are uncommon. The main benefit of C3 plants in cold climates is that they are adapted to thrive when temperatures are cooler, which allows them to outcompete other plant types, such as C4 or CAM plants, in these regions.

KEY CHARACTERISTICS OF C3 PLANTS

Cool-Season Growth: They prefer cool temperatures and can handle early spring and late fall frosts.

Long Daylight: C3 plants benefit from long daylight hours, often seen in northern regions during summer.

High Water Needs: While they do well in cooler climates, they still require adequate moisture for optimal growth.

Vulnerable to Heat Stress: They struggle in hot, dry climates, making them unsuitable for extreme summer heat in tropical or desert regions.

Best C3 Plants for Northern Cold Climates

Here's a list of some of the best C3 plants to grow in northern, cold climates. These plants are naturally suited for cool weather, frost tolerance, and shorter growing seasons.

1. LEAFY GREENS

Leafy greens are the bread and butter of cold-climate gardens. They are fast-growing, tolerate frost, and can be planted early in the season.

Spinach: Spinach thrives in cool weather, and it can even tolerate light frosts. It grows quickly and can be harvested early in the season.

Lettuce: Varieties such as butterhead, romaine, and leaf lettuce flourish in northern climates. These varieties grow quickly and can handle cooler temperatures.

Kale: This superfood thrives in cold weather and even tastes sweeter after a frost. Kale can be grown from early spring to late fall.

Arugula: A fast-growing green that thrives in cool weather, perfect for early spring planting.

2. ROOT VEGETABLES

Root vegetables are well-suited to colder climates because they grow underground, where the soil stays cooler and more stable than above-ground.

Carrots: Carrots grow best in cool weather and can tolerate light frosts. Choose early varieties for a quicker harvest.

Beets: Beets thrive in cooler conditions and can even be sown early in the spring. Their greens are also edible and nutrient-rich.

Radishes: Radishes are one of the quickest-growing root vegetables and are perfect for cold climates. They can be grown early in the season or even in succession plantings.

Turnips: Another great root vegetable, turnips are hardy and can survive in temperatures as low as 20°F (-6°C) once mature.

3. BRASSICAS (CABBAGE FAMILY)

Brassicas are cold-hardy plants that are excellent choices for northern climates. Many of them can withstand light frosts and mature well in cooler temperatures.

Cabbage: Cabbage varieties such as green, red, and Savoy are excellent for northern climates. They grow best in the cooler months and can be harvested before the first hard frost.

Broccoli: Cold-hardy varieties of broccoli can be grown in northern climates. Start them early for a summer harvest.

Cauliflower: Like broccoli, cauliflower thrives in cool temperatures. They need ample time to grow, but cold-hardy varieties can be harvested in late summer.

Brussels Sprouts: Known for their tolerance of frost, Brussels sprouts actually taste sweeter after a frost, making them an excellent choice for northern climates.

4. PEAS AND BEANS

Certain legumes thrive in the cooler temperatures of northern climates, providing a healthy source of protein and nutrients.

Peas: Both snow peas and sugar snap peas thrive in cool weather. They can be planted early and will produce abundant yields throughout the spring and early summer.

Fava Beans: Fava beans are one of the hardiest legumes, able to withstand cold temperatures and frost. They are excellent for northern climates and provide a high-protein food source.

5. HERBS

Cold-hardy herbs can add flavor to your meals and thrive in northern climates with a little care.

Parsley: Parsley is a biennial herb that grows well in cooler temperatures and can tolerate light frosts.

Chives: Chives thrive in cool climates and can even be harvested in early spring as soon as the soil is workable.

Thyme: Thyme is a hardy perennial herb that will grow in the cold, surviving well into the fall and winter with some protection.

6. SMALL FRUITS

Although fruit production is limited in cold climates, there are a few hardy options that can withstand freezing temperatures.

Strawberries: Varieties like Alpine or June-bearing strawberries are cold-tolerant and can produce fruit in northern climates with proper care.

Raspberries: Cold-hardy varieties of raspberries, such as Boyne and Heritage, are perfect for northern growing zones and can withstand frosty conditions.

Currants and Gooseberries: These hardy berries are well-suited for colder climates and can provide excellent yields in the right conditions.

Growing and Caring for C3 Plants

1. Start Early, Harvest Early

C3 plants tend to grow quickly, which is especially beneficial in short growing seasons. Start seeds indoors or in a greenhouse, or use cold frames or cloches to extend the growing season and protect plants from frost. Many C3 plants can be harvested early or in succession, allowing for multiple rounds of growth.

2. Soil Preparation

Cool-climate plants thrive in well-drained, fertile soil. Amend your soil with compost and organic matter to ensure the soil retains moisture but doesn't become waterlogged. Adding lime can help raise the pH in acidic soils, which is especially beneficial for root crops and brassicas.

3. Maximize Sunlight

Northern climates tend to have long daylight hours in the summer but shorter growing seasons. Maximize sunlight exposure by planting C3 crops in full sun or in areas that receive the most direct light. Raised beds or containers can also help warm up the soil early in the season, encouraging faster growth.

4. Protect from Frost

While C3 plants are frost-tolerant, there's a limit. To extend your growing season and protect tender crops from early frosts, use row covers, hoop houses, or greenhouses to maintain a warm microclimate around your plants.

C3 plants are your go-to choice for northern cold climates. Their natural ability to thrive in cooler temperatures and longer daylight hours makes them perfect candidates for your homestead garden. With the right care, C3 plants will produce bountiful harvests in the shortest growing seasons, from leafy greens and root vegetables to hardy brassicas and legumes.

By embracing these cool-season crops, you'll maximize your food production while making the most of the natural advantages your cold climate offers.

C4 Plants for Arid and Tropical Climates

WHAT ARE C4 PLANTS?

C4 plants use a specialized form of photosynthesis known as C4 photosynthesis, which helps them efficiently handle high temperatures, intense sunlight, and low water availability. In simple terms, C4 plants have an advantage in hot and dry climates because they can fix carbon more efficiently under stress. Unlike C3 plants, which open their stomata more often to capture carbon dioxide, C4 plants open their stomata less, conserving water. They are thus more efficient at photosynthesis in both high heat and low moisture environments.

WHY ARE C4 PLANTS IDEAL FOR ARID AND TROPICAL CLIMATES?

Water Efficiency: C4 plants are highly water-efficient, making them perfect for areas with inconsistent rainfall, limited water availability, or high evaporation rates.

Heat Tolerance: These plants have adapted to thrive in hot climates, making them resilient to extreme temperatures and drought conditions, both of which are common in arid and tropical regions.

Faster Growth: Many C4 plants grow quickly, which can help maximize productivity in climates where the growing season might be short or irregular.

Common C4 Plants for Arid and Tropical Climates

C4 plants are quite diverse. They encompass grains, grasses, and other types of plants, many of which are staples in both arid and tropical environments.

Here are some of the best-known C4 plants for these regions:

1. CORN (ZEA MAYS)

Suitability for Arid and Tropical Climates: Corn is one of the most well-known C4 plants, and it thrives in regions with plenty of sunlight and warmth. It is especially suited for arid climates if irrigated properly and can also flourish in tropical regions during the dry season.

Water Needs: While corn is relatively water-efficient compared to C3 plants, it does require regular irrigation during dry periods to reach optimal yield.

Growing Tips:

Plant corn in full sun where it gets at least 6–8 hours of direct light.

Corn benefits from well-drained soil with good organic matter content.

Space plants adequately to ensure good airflow, as this helps to prevent fungal diseases, especially in tropical climates.

2. SORGHUM (SORGHUM BICOLOR)

Suitability for Arid and Tropical Climates: Sorghum is a drought-tolerant, heat-loving C4 plant well-suited for both arid and tropical climates. It is often used as a grain, fodder, and even for biofuels.

Water Needs: Sorghum requires less water than many other grains, making it ideal for arid climates. It's also quite adaptable in tropical regions, where it can withstand both dry periods and short heavy rains.

Growing Tips:

Sorghum thrives in well-drained, sandy loam soils.

It requires a long growing season (about 90-120 days) and plenty of sunlight.

In arid climates, consider planting it in the early

spring to avoid extreme heat during flowering.

3. MILLET (PENNISETUM GLAUCUM, PANICUM MILIACEUM)

Suitability for Arid and Tropical Climates: Millet is another highly drought-tolerant C4 plant that thrives in arid and semi-arid regions. It also grows well in the tropics, where it handles the intense sun and periodic rainfall with ease.

Water Needs: Millet needs less water than other crops, making it ideal for dry climates. It is well-suited to areas with unreliable rainfall or those facing periods of drought.

Growing Tips:

Millet requires loose, well-drained soil.

It's tolerant of poor soils and can even grow in soils with low fertility.

Millet seeds should be planted shallowly and in rows to maximize the spacing for air circulation and light exposure.

4. SUGARCANE (SACCHARUM OFFICINARUM)

Suitability for Arid and Tropical Climates: Sugarcane is a classic C4 crop that thrives in both tropical and arid regions, though it requires a steady water supply for optimal sugar production. It is widely grown in tropical climates but can also be found in arid regions with irrigation systems in place.

Water Needs: Sugarcane requires substantial irrigation during its growing season, but its C4 photosynthesis allows it to perform better in hot, dry conditions than many C3 crops.

Growing Tips:

Sugarcane needs rich, well-drained soils with plenty of organic matter.

Plant it in full sun with temperatures consistently above 77°F (25°C) for optimal growth.

It's best suited to areas where irrigation can be maintained throughout the growing period.

5. TEFF (ERAGROSTIS TEF)

Suitability for Arid and Tropical Climates: Teff is a highly drought-resistant C4 plant originally from Ethiopia, where it thrives in both arid and semi-arid climates. It is well-suited for tropical regions as well, where it can be grown in the rainy season or during periods of dry weather.

Water Needs: Teff requires relatively low amounts of water compared to other grains, making it perfect for dry conditions.

Growing Tips:

Plant teff in well-drained soils with a slightly acidic pH.

Teff does well in raised beds or small plots.
Its tiny seeds need to be planted shallowly—about 1/8 inch deep.

6. BAMBOO (VARIOUS SPECIES)

Suitability for Arid and Tropical Climates: Bamboo is a C4 plant that grows quickly in tropical regions and can also tolerate drier arid areas when given adequate irrigation.

Water Needs: Bamboo needs moderate watering but can tolerate drought once established.

Growing Tips:

Bamboo thrives in well-drained soil with good moisture retention.

Once established, bamboo can spread quickly, so containment or control may be necessary to prevent it from overtaking your garden.

GENERAL TIPS:

1. Soil Management

Well-Drained Soil: Most C4 plants perform best in well-drained, loose soil. Heavy, compacted soils can reduce water infiltration, especially during the hot, dry season.

Soil Fertility: C4 plants are generally more tolerant of poor soil conditions than C3 plants, but amending the soil with compost and organic matter can improve growth, particularly for crops like corn and sorghum.

2. Watering Efficiency

While C4 plants are more water-efficient than C3 plants, they still need regular irrigation, especially during dry spells.

Drip irrigation or soaker hoses are great ways to ensure that water is delivered directly to the root zone, minimizing evaporation and ensuring efficient water use.

3. Heat Management

While C4 plants are heat-tolerant, it's still important to protect them from the extreme midday sun in arid climates. Using shade nets or row covers can help regulate the temperature around the plants.

4. Pest and Disease Management

In both arid and tropical climates, pests like aphids, grasshoppers, and spider mites can affect C4 crops. Using natural methods like predatory insects, neem oil, or companion planting can help keep pests in check.

C4 plants are the champions of hot, dry environments, making them ideal for both arid and tropical climates. Their ability to thrive in extreme temperatures while conserving water allows you to grow crops that would otherwise struggle in these regions. From corn and sorghum to sugarcane and teff, these plants not only provide a reliable source of food but also offer incredible drought tolerance. By understanding their unique needs and tailoring your growing practices, you can make the most of these hardy, efficient plants in your homestead, creating a sustainable food supply even in the harshest climates.

Homesteading in alignment with your climate isn't just about adapting—it's about thriving. When you understand the unique strengths and challenges of your environment, you unlock the full potential of your land.

CHAPTER TWO: HOMESTEAD DESIGN

Designing an advanced homestead isn't just about making the most of your land—it's about creating a resilient, self-sustaining ecosystem tailored to your goals. In this chapter, we'll go far beyond the basics to cover zoning, structural positioning, permaculture principles, companion planting, and land use efficiency. Think of this as your homestead's master plan, blending practicality and vision into a thriving, sustainable future.

Advanced Zoning Techniques—Strategic Land Division

Zoning your homestead is like organizing a symphony. Each "zone" has a role to play, ensuring that your efforts harmonize into a productive, efficient whole. But in advanced homesteading, zoning isn't just theoretical—it's dynamic, adaptable, and designed to evolve with your needs.

MAPPING YOUR HOMESTEAD

Start with a Base Map: Use tools like Google Earth, a drone, or even graph paper to sketch your land. Include natural features (elevation changes, trees, water sources) and existing structures.

Pro Tip: Overlay a sun path diagram to track light exposure throughout the year.

Deep Dive into Zoning

Zones aren't static—they reflect your movement patterns and priorities. Let's break it down with advanced applications.

Zone 1: High-Frequency Use Areas

Zone 1 is the beating heart of your homestead—a place where convenience meets productivity. It includes spaces and features you'll visit daily or even multiple times a day. Whether you're harvesting herbs for tonight's dinner, checking on the compost bin, or collecting eggs, Zone 1 is designed to maximize efficiency while keeping you inspired.

KITCHEN GARDEN MASTERY

Think of your kitchen garden as a living pantry.

Perennial Herbs in Pathways: Use herbs like rosemary, thyme, sage, and oregano to create natural, aromatic pathways. Interplant with edible flowers like nasturtiums and calendula for bursts of color and additional culinary use.

Vertical Gardening: Install trellises or wall-mounted planters for high-yield crops like cherry tomatoes, pole beans, and cucumbers. This saves space and creates a lush, layered look.

Edible Ground Covers: Incorporate creeping thyme or strawberries as living mulch to suppress weeds and provide a harvestable bonus.

COLD FRAMES FOR YEAR-ROUND GREENS

Take advantage of cold frames to extend your growing season. Use them for crops like arugula, spinach, mizuna, and claytonia in colder months.

Tip: Build hinged or sliding cold frame covers to make ventilation and harvesting easier. For insulation, line the sides with straw bales during winter.

MICROCLIMATES

Leverage structures, walls, and hedges to create microclimates. Plant heat-loving herbs and vegetables (like basil and cherry tomatoes) near sun-facing walls for earlier harvests.

Tip: Position shade-loving plants, such as lettuce or cilantro, behind taller crops to protect them from excessive sun.

WATER MANAGEMENT IN ZONE 1

Rain Barrel Systems

Installing rain barrels in Zone 1 gives you immediate access to water without running hoses or carrying buckets long distances. Set up a series of linked barrels to increase capacity and efficiency. Include a first-flush diverter to filter out debris and contaminants from the initial rain runoff.

Tip: Add spigots at varying heights for watering cans or drip systems.

Drip Irrigation on a Timer

For your most frequently watered beds, install drip irrigation systems with timers. This ensures plants stay hydrated while freeing you up for other tasks.

HIGH-FREQUENCY COMPOSTING

Zone 1 is an ideal spot for your kitchen waste composting system.

Compost Tumblers: These are great for quick-turnaround compost, allowing you to process small amounts of kitchen scraps efficiently.

Vermicomposting: A worm bin in Zone 1 can handle daily food scraps while producing nutrient-rich vermicast. Keep it close for easy access during food prep.

Tip: Incorporate biochar into your compost to

improve microbial activity and increase nutrient retention in your soil.

LIGHTING FOR EVENING TASKS

For those moments when you find yourself gardening or working late into the evening, install solar-powered path lights along garden pathways to ensure safety and visibility.

Consider motion-sensor LED lights near frequently visited areas like the compost bin or chicken coop.

Wildlife and Pest Management

Zone 1's proximity to the house means it's critical to balance attracting beneficial wildlife and deterring pests. Plant insectary strips with yarrow, dill, and fennel to attract pollinators and predatory insects. Use borders of garlic, onions, or marigolds to deter deer, rabbits, and other unwanted visitors.

Tip: Place sticky traps or pheromone traps for pests like whiteflies or aphids to catch early infestations.

INTEGRATED OUTDOOR LIVING SPACES

Zone 1 isn't just functional—it's also the perfect place to relax and recharge.

Install a small patio or seating area to enjoy the fruits of your labor. Incorporate a small firepit or outdoor oven to cook meals using ingredients harvested steps away.

Tip: Use shade sails or pergolas covered with grapevines or kiwis for a natural, cooling canopy.

Maximizing Efficiency Through Layout

Tool Storage: Keep a small tool shed or wall-mounted tool rack in Zone 1 to avoid unnecessary trips to your main storage.

Planting in Succession: Rotate fast-growing crops like radishes, salad greens, and carrots in raised beds to keep the kitchen garden perpetually productive.

Raised Beds with Access Paths: Design your beds so you can reach every corner without stepping into the planting area. A 4-foot width is ideal.

Zone 1 is your daily partner on the homestead. By integrating thoughtful design, efficient systems, and a touch of creativity, this high-frequency use area becomes both a workhorse and a source of joy.

Zone 2: Perennial and Pollinator Paradise

Zone 2 is the powerhouse of long-term productivity on your homestead. This zone strikes a perfect balance between semi-regular maintenance and the beauty of self-sustaining systems. Think of it as your "set it and support it" zone, where perennials, orchards, and pollinator habitats thrive, working in harmony to create a diverse and resilient ecosystem.

ADVANCED ORCHARD GUILD DESIGN

An orchard guild is more than just a cluster of fruit trees—it's an interdependent system where every plant plays a role in health, productivity, and resilience.

NITROGEN-FIXING CHAMPION

Pair fruit trees with nitrogen-fixing plants like Siberian pea shrub, goumi, or autumn olive to enrich the soil naturally.

Tip: Use nitrogen-fixers as living mulch—trim them regularly and drop the cuttings around your fruit trees to recycle nutrients.

DYNAMIC ACCUMULATORS

Integrate plants like comfrey and yarrow that mine deep soil nutrients and make them available to surface plants.

Tip: Plant comfrey at the drip line of your fruit trees. Use its leaves as a chop-and-drop mulch to deliver potassium and phosphorus.

LIVING PEST CONTROL

Plant garlic, chives, or tansy around the base of trees to repel pests like aphids and borers.

Tip: Interplant with marigolds and nasturtiums, which deter nematodes and attract beneficial predatory insects like lacewings.

GROUND COVERS FOR WEED SUPPRESSION

Replace bare soil with creeping thyme, clover, or strawberries to reduce weeding, retain moisture, and provide additional harvests.

Example Guild: Surround a plum tree with comfrey for nutrient cycling, goumi for nitrogen-fixing, garlic for pest control, and white clover for living mulch.

POLLINATOR PATCHES

Pollinators are the lifeblood of a productive homestead. Establishing dedicated pollinator habitats in Zone 2 ensures robust fruit set and creates a biodiversity haven.

WILDFLOWER MEADOWS

Plant a mix of native wildflowers like echinacea, milkweed, and black-eyed Susan to support bees, butterflies, and other pollinators.

Tip: Include staggered bloom times in your seed

mix to ensure flowers are available from early spring to late fall.

PERENNIAL POWERHOUSES

Add perennials that pull double duty by supporting pollinators and offering medicinal or culinary benefits.

Examples:

Lavender: Attracts bees and can be used for teas, oils, and sachets.

Fennel: Hosts beneficial insects like hoverflies and ladybugs.

Borage: A favorite for bees and produces edible flowers.

POLLINATOR-FRIENDLY INFRASTRUCTURE

Install bee hotels, butterfly houses, or even a small hive in this zone. Keep it far enough from heavy traffic but close enough for easy access.

FOOD FOREST EXPANSION

Zone 2 is ideal for a food forest, where every layer of vegetation contributes to productivity and resilience.

Canopy Layer: Fruit and nut trees like apple, pear, or hazelnut form the upper layer.

Understory Layer: Shrubs like blueberries, currants, or elderberries thrive in the dappled shade.

Herbaceous Layer: Perennials like rhubarb, mint, or lemon balm offer culinary and medicinal harvests.

Ground Cover Layer: Plant low-growing crops like creeping oregano or wild strawberries.

Root Layer: Include nutrient-rich root crops like Jerusalem artichokes or garlic for underground yields.

Vine Layer: Use trellises or tree trunks for climbers like grapes, kiwis, or passionfruit.

INTEGRATED WATER MANAGEMENT

Zone 2 is an excellent location to incorporate water features and passive irrigation systems.

Swales and Berms

Capture rainwater by digging swales (shallow trenches on contour) and planting along the berms (raised soil beside the swales). This technique slows water runoff and channels it to your perennial plants. Advanced Tip: Line swales with mulch or plant clover to stabilize soil and boost nitrogen.

Pond Integration

Small ponds or wetlands not only provide water for plants and livestock but also attract beneficial wildlife like frogs and dragonflies that help control pests.

Example: Install a solar-powered pump to circulate water in your pond and irrigate nearby plants.

Wildlife Management

Zone 2 provides a perfect balance of welcoming beneficial wildlife while protecting your crops.

Bird Attractors

Place birdhouses and feeders strategically to invite pest-controlling species like wrens and bluebirds. Advanced Tip: Grow sunflowers or millet in this zone as a natural birdseed supply.

Wildlife Exclusion

Use fencing or natural barriers like dense hedgerows to deter larger animals like deer or rabbits. For added protection, grow thorny plants like blackberries along the perimeter.

DIVERSIFIED HARVESTS

Zone 2 is your homestead's investment portfolio—a place where perennial crops offer consistent returns year after year.

Nut Trees: Consider high-yield options like chestnuts, pecans, or walnuts.

Berry Crops: Blackberries, raspberries, and blueberries are low-maintenance and incredibly productive.

Medicinal Plants: Cultivate perennials like valerian, St. John's wort, or ginseng for advanced herbal remedies.

ZONE 2 EFFICIENCY

Pathways: Design wide, mulched pathways that can accommodate wheelbarrows for easy maintenance and harvest.

Perimeter Planting: Use hedges like elderberries or hazelnuts along Zone 2's edge for dual-purpose windbreaks and harvests.

Zone Integration: Locate Zone 2 close to Zone 1 for efficient transitions, ensuring daily or weekly visits are convenient.

Zone 2 is where the magic of long-term productivity begins. By designing resilient systems that require less daily input, you create a homestead that works for you. Whether you're harvesting sun-ripened fruit, admiring pollinators at work, or chopping comfrey for mulch, this zone epitomizes the balance between effort and abundance.

Zone 3: Livestock and Pasture Management

Zone 3 is the beating heart of your homestead's larger-scale, long-term food production and land stewardship. This zone focuses on sustainable and efficient livestock care, pasture management, and integrating productive ecosystems like silvopasture. The goal? A harmonious balance where your animals thrive, your land improves over time, and your workload is streamlined.

ROTATIONAL GRAZING SYSTEMS

Rotational grazing is more than just dividing a pasture into sections. It's about creating a dynamic, regenerative system that boosts soil health, reduces feed costs, and keeps your animals healthier.

STEP-BY-STEP SETUP

Divide Pastures: Use portable electric fencing to create paddocks of manageable size. Each paddock should provide enough forage for 1–7 days.

Timing is Key: Allow grazed paddocks to rest for 30–60 days, depending on the season and growth rate of your grasses.

Water Access: Use portable water troughs or permanent water points within easy reach of each paddock.

Animal Integration

Combine species for complementary grazing patterns. For example, cattle graze tall grasses, sheep target lower growth and weeds and goats browse woody plants and shrubs.

Tip: Follow large livestock with chickens. As they scratch through droppings, they reduce pest loads (like fly larvae) and spread nutrients evenly across the paddock. This reduces the need for additional pest control or fertilization.

SILVOPASTURE: THE FUTURE OF SUSTAINABLE GRAZING

Silvopasture integrates trees and shrubs into grazing lands, creating a multifunctional landscape that benefits livestock, wildlife, and the environment.

BENEFITS OF SILVOPASTURE

Shade: Keeps animals cool and reduces heat stress, improving weight gain and milk production.

Forage: Trees like mulberries, carob, or locust provide edible leaves and fruits for livestock.

Soil Health: Tree roots reduce erosion and improve water retention.

TREE SELECTION FOR SILVOPASTURE

Choose trees that serve multiple purposes:

Mulberry: High-protein leaves for livestock and fruit for human consumption.

Black Locust: Nitrogen-fixing tree with durable wood for posts and fuel.

Willow: Excellent fodder, especially during droughts.

Persimmon: Provides fruit for animals and humans alike.

Implementation: Space trees widely enough (20–40 feet apart) to allow sunlight to reach pasture grasses. Use tree guards to protect young saplings from browsing animals.

MANAGED GRAZING FOR SPECIFIC LIVESTOCK

Each species has unique grazing habits and needs. Tailoring your approach ensures healthier animals and optimized land use.

Cattle

Large grazers that thrive on grasses. Rotate

frequently to prevent overgrazing, which can lead to soil compaction.

Tip: Introduce a "leader-follower" system where cattle graze first, followed by smaller species like sheep to target weeds and lower growth.

Goats

Natural browsers, goats excel at clearing woody plants, shrubs, and invasive weeds.

Tip: Use goats to clear overgrown areas before converting them into silvopasture or rotational grazing paddocks.

Pigs

Pigs are excellent for rooting and tilling. Use them to prep land for reseeding or to turn compost piles. *Tip*: Rotate pigs through wooded areas to forage on acorns, which add flavor to pork.

Sheep

Ideal for weed control and grazing on smaller plots. They prefer short, tender grasses.

Tip: Incorporate sheep into vineyards or orchards—they graze on weeds without damaging tree bark or vines (with proper protection).

ADVANCED PASTURE MANAGEMENT TECHNIQUES

Healthy pastures are the foundation of Zone 3. Implement advanced strategies to maintain lush, productive grazing land.

Overseeding and Reseeding

Reseed pastures annually with a mix of grasses and legumes like alfalfa, red clover, or ryegrass. These plants improve forage quality, boost nitrogen levels, and prevent erosion.

Tip: Use a no-till drill to plant seeds directly into existing pastures without disturbing the soil structure.

Pasture Rest and Recovery
Avoid continuous grazing, which depletes grasses and weakens root systems. Rest periods of 30–60 days allow plants to regrow fully.

SOIL TESTING AND AMENDMENTS

Test your soil annually to monitor pH and nutrient levels. Amend with lime to correct acidity or with composted manure to add organic matter.

Tip: Plant cover crops like buckwheat or field peas during off-seasons to suppress weeds and enrich the soil.

WATER MANAGEMENT IN ZONE 3

Effective water use is crucial for large-scale livestock and pasture management.

Ponds and Dams
Build or enhance ponds to serve as reliable water sources for livestock. Surround them with vegetation to reduce evaporation and filter runoff.

Tip: Use a solar-powered pump to distribute water from the pond to remote paddocks.

Portable Water Systems
Use mobile water tanks and troughs for flexibility in rotational grazing setups. This reduces soil compaction around permanent water points.

Pasture Drainage
Prevent waterlogging by installing French drains or surface ditches in low-lying areas.

WILDLIFE INTEGRATION AND MANAGEMENT

Zone 3 naturally attracts wildlife. Strive for balance by welcoming beneficial species while protecting your resources.

Beneficial Wildlife

Encourage predatory birds, like hawks or owls, by installing perches to control rodents in grazing areas.

Tip: Allow hedgerows to grow along the edges of Zone 3 to provide habitat for pollinators, birds, and beneficial insects.

Fencing Solutions

Use electric fencing to deter predators like coyotes or foxes. Solar-powered chargers make it easy to electrify remote fencing.

DIVERSIFYING ZONE 3

Zone 3 isn't just for grazing—it can also provide additional yields with minimal extra effort.

Honey and Pollination

Place beehives near silvopasture or wildflower areas to boost forage production and harvest honey.

Tip: Choose hive designs like top-bar hives, which are easier to inspect and maintain in remote locations.

Timber and Firewood

Harvest fast-growing trees like poplar or black locust for firewood or lumber while maintaining their benefits as shade providers.

Edible Fodder Crops

Grow high-yield crops like mangels (fodder beets) or alfalfa to supplement winter feed for livestock.

Zone 3 represents the productive, wide-open spaces of your homestead, where livestock, pasture, and ecosystems work together. With advanced techniques like rotational grazing, silvopasture, and multi-species integration, this zone is your key to sustainable and profitable livestock management.

By investing in the health of your animals and land, you're building a self-reliant future that thrives year after year.

Zone 4: Forest and Resource Zones

Zone 4 is where the wild meets the cultivated, blending natural ecosystems with strategic management to provide an endless supply of resources. This is your homestead's long-term investment area, offering timber, foraged foods, and medicinal plants, all while enhancing biodiversity and supporting a healthy environment. Proper planning and sustainable practices ensure that your forest and resource zones remain productive for generations.

TIMBER HARVEST AND MANAGEMENT

Managing your forested areas isn't just about cutting trees—it's about fostering a renewable and balanced system. Timber provides firewood, construction materials, and even income when managed wisely.

Selective Logging
Identify and mark mature or overcrowded trees for harvest. Leave younger, healthy trees to grow. Use low-impact methods like a chainsaw with a winch or small-scale forestry equipment to minimize soil disruption. Maintain a mix of species to ensure ecological balance.

COPPICING AND POLLARDING

Coppicing involves cutting trees like willow, poplar, or hazel to ground level, encouraging regrowth from the stump. This method provides a renewable wood source for firewood, fencing, or crafts.

Pollarding is similar but involves cutting trees

higher up, which protects the regrowth from browsing animals.

TIMBER AS INCOME

High-value hardwoods like oak, maple, or walnut can be selectively harvested for specialty markets, such as furniture-making or veneer.

Tip: Mill your lumber on-site with a portable sawmill to increase its value and use it for homestead projects.

Forest Maintenance

Thin overcrowded stands to improve light penetration and encourage the growth of understory plants. Remove invasive species like kudzu or honeysuckle to protect native biodiversity.

FORAGING: WILD FOOD AND MEDICINE

Zone 4 is a treasure trove of wild edibles, medicinal plants, and fungi. Learning to identify, harvest, and process these resources connects you deeply to your land.

Wild Edibles

Mushrooms: Cultivate gourmet varieties like shiitake or oyster mushrooms on logs in shaded areas, or forage for wild types like morels and chanterelles.

Berries: Encourage brambles like blackberries, raspberries, and elderberries by managing light levels and pruning old canes.

Nuts: Look for species like black walnut, chestnut, or pecan for both human consumption and livestock feed.

Medicinal Plants

Wild herbs like ginseng, goldenseal, or yarrow can

be sustainably harvested and processed into tinctures, teas, or salves.

Tip: Create a small "wild pharmacy" by growing or encouraging medicinal plants in designated areas.

Sustainable Foraging Practices

Harvest no more than 20–30% of a given plant or fungi population to allow natural regeneration. Rotate foraging areas yearly to avoid depleting resources.

BIODIVERSITY AND WILDLIFE MANAGEMENT

Zone 4 is a haven for wildlife, which plays a critical role in maintaining ecological balance. By encouraging biodiversity, you can protect your forest resources while benefiting from natural pest control and pollination.

Wildlife Corridors

Create undisturbed areas within Zone 4 to serve as habitats for deer, foxes, and small mammals. Install nest boxes for birds like owls, which help control rodent populations.

Edible Wildlife Support

Plant mast-producing trees like oaks (acorns), hickories (nuts), or fruit-bearing shrubs to attract game species like turkeys or deer.

Forest Edge Management

The edge of Zone 4 can be planted with flowering shrubs like elderberry or dogwood, which support pollinators and provide food for humans and animals.

INTEGRATING AGROFORESTRY PRACTICES

Agroforestry in Zone 4 blends forestry and

agriculture to create multifunctional systems that boost productivity and ecological health.

Alley Cropping

Plant rows of trees or shrubs with wide spaces in between for crops like berries, herbs, or even forage for livestock.

Example: Rows of black locust trees with strips of perennial herbs like lavender or oregano in between.

Forest Gardening

Mimic natural forest layers with plantings of nut trees (overstory), fruit trees (midstory), berry bushes (understory), and ground covers like strawberries or clover.

ADVANCED WATER RESOURCE MANAGEMENT

Zone 4 often contains natural water features like streams, ponds, or wetlands. Managing these resources responsibly enhances the health of your forest and supports its productivity.

POND AND STREAM CARE

Protect riparian zones (the areas along watercourses) by leaving a buffer of natural vegetation. Install sediment traps or small dams to reduce erosion and improve water quality.

RAINWATER HARVESTING IN ZONE 4

Use contour bunds or swales to direct water into your forested areas, enhancing tree growth and reducing runoff.

FIREWOOD MANAGEMENT

Firewood is a cornerstone resource for many

homesteaders. Zone 4 can provide a steady, renewable supply when harvested responsibly.

Cutting and Storing Firewood

Focus on fast-growing species like poplar or pine for regular harvesting, and save slower-growing hardwoods like oak for long-term needs.

Store split wood in covered stacks, allowing it to season for 6–12 months to reduce moisture content and improve burn efficiency.

Woodlot Improvements

Plant additional fast-growing trees in underutilized areas of Zone 4 to ensure a future firewood supply.

WILDCRAFTING AND SUSTAINABLE CRAFTING MATERIALS

Zone 4 provides raw materials for countless homestead projects, from crafting to construction.

Basketry and Weaving

Harvest willow, cattail, or reeds for making baskets, mats, or decorative items.

Tip: Create a willow coppice dedicated to basketry supplies.

Natural Building Materials

Harvest small-diameter trees for poles, posts, or wattle fencing.

Use bark or moss for roof shingles or natural insulation.

Wildcrafting Projects

Make dyes from plants like black walnut (brown) or pokeweed (pink). Craft homemade furniture from selectively harvested timber.

Zone 4 is a place where the untamed landscape becomes a source of endless potential. Whether you're managing timber, foraging for wild edibles, or nurturing wildlife habitats, this zone teaches you the art of working with nature rather than against it.

By thoughtfully integrating practices like agroforestry, sustainable harvesting, and wildcrafting, you can transform your forested areas into a rich, regenerative resource for your homestead.

Zone 5: The Wild Zone

Zone 5 is the untouched, unmanaged heart of your homestead—a sanctuary for wildlife and a space where nature is allowed to flourish undisturbed. While it may seem like a passive area, this zone plays an essential role in supporting the overall health and balance of your homestead. By observing and learning from this space, you can gain insights into natural ecosystems that directly benefit your managed zones.

Think of Zone 5 as your outdoor laboratory and ecological safety net.

THE ROLE OF BIODIVERSITY IN ZONE 5

Maintaining a wild zone encourages biodiversity, which provides countless benefits to your homestead. Native plants, animals, and insects work in harmony to regulate pests, pollinate crops, and build soil health.

Native Wildlife Support

Birds like hawks, owls, and kestrels are natural rodent predators. Install perches or nest boxes in Zone 5 to attract these beneficial species.

Mammals like foxes and weasels also contribute to rodent control and can be encouraged with undisturbed brush piles.

Insect Allies

Predatory insects like dragonflies and ladybugs thrive in wild zones, keeping pest populations in check.

Maintain areas with native flowering plants to provide food and shelter for pollinators like bees and butterflies.

Aquatic Biodiversity

If your Zone 5 includes wetlands, streams, or ponds, protect riparian vegetation to support aquatic species like frogs, which control insects.

OBSERVATION AS A TOOL IN ZONE 5

Zone 5 offers an incredible opportunity to learn from nature's systems and cycles. By observing patterns, you can apply this knowledge to other zones on your homestead.

Wildlife Behavior

Monitor the presence of specific birds, mammals, or insects as indicators of ecosystem health. For example, seeing a variety of pollinators might suggest a thriving native plant community.

Track seasonal migrations or nesting habits to understand the rhythms of your local wildlife.

Natural Patterns

Study water flow during heavy rains to identify natural drainage patterns and prevent erosion in other zones.

Observe how native plants adapt to drought, pests, or soil conditions and mimic their strategies in your managed areas.

NATIVE PLANT CONSERVATION

Zone 5 is the perfect place to nurture native plants, which are adapted to your local climate and soil. These plants often require no maintenance, making them ideal for this unmanaged area.

BENEFITS OF NATIVE PLANTS

Deep root systems stabilize soil and improve water retention, preventing erosion.

Native plants provide food and habitat for local wildlife, including birds, bees, and butterflies.

HOW TO ENCOURAGE NATIVE PLANT GROWTH

Remove invasive species like kudzu or Japanese honeysuckle to give native plants a chance to thrive. Spread seeds or plant native shrubs in areas where erosion or degradation has occurred.

WILDCRAFTING IN ZONE 5

While Zone 5 is primarily left untouched, it can still provide occasional resources for the homestead. Sustainable wildcrafting ensures you're harvesting responsibly without depleting the ecosystem.

Edibles and Medicinals

Gather small amounts of wild herbs like stinging nettle or dandelion for teas and tonics.

Harvest berries or nuts sparingly, leaving enough for wildlife to thrive.

Natural Materials

Collect fallen branches or driftwood for crafting or firewood, but avoid disturbing live trees or habitats.

Use pine needles or moss as natural mulch or for

crafting projects like wreaths.

Pro Tip: Rotate your wildcrafting areas and leave ample time for ecosystems to regenerate between harvests.

CREATING WILDLIFE HABITATS

Enhancing Zone 5 as a wildlife sanctuary benefits not only your local ecosystem but also your managed zones by attracting beneficial species.

Habitat Structures

Build brush piles from fallen branches to provide shelter for small mammals and insects. Install logs or stumps as habitat for fungi, amphibians, and insects. Add small ponds or depressions for frogs, toads, and other water-loving creatures.

Encouraging Nesting

Leave old trees standing when safe, as they provide nesting sites for birds and bats. Plant dense thickets or hedgerows to give ground-nesting birds a safe space.

FIREBREAKS AND LAND STEWARDSHIP

In dry climates or areas prone to wildfires, Zone 5 can also serve as a fire buffer when managed thoughtfully.

Firebreaks

Maintain low vegetation and clear debris along the borders of Zone 5 to create a firebreak that protects your homestead. Use controlled burns (where permitted) to manage excess plant material and rejuvenate growth.

Stewardship Without Overmanagement

Strike a balance between encouraging natural

processes and preventing potential hazards like invasive plants or excessive deadwood.

Monitor Zone 5 periodically to ensure it remains a healthy and beneficial part of your ecosystem.

SPIRITUAL AND PERSONAL RENEWAL

Beyond its ecological and practical benefits, Zone 5 serves as a space for personal connection with nature. It's a place to retreat, reflect, and recharge.

Nature Walks

Use trails or pathways for quiet walks where you can observe wildlife, plants, and the changing seasons.

Tip: Bring a notebook to document what you see—your Zone 5 observations can inspire garden plans, art projects, or even foraging guides.

Mindfulness and Meditation

Dedicate a spot within Zone 5 for sitting quietly or meditating. The undisturbed natural sounds and sights can be grounding and restorative. Install a simple bench or stone seat to create a welcoming meditation space.

Zone 5 embodies the spirit of working with nature instead of against it. By leaving this zone largely untouched, you support a thriving ecosystem that benefits every aspect of your homestead. Through observation, occasional wildcrafting, and mindful stewardship, Zone 5 becomes both a resource and a reminder of nature's resilience and abundance. It's a space where you can step back, learn, and let the wild inspire you.

Positioning Structures for Efficiency

Structures are more than just functional—they can shape your daily routine, reduce energy use, and improve workflow. Advanced placement strategies make all the difference.

OPTIMIZE FOR SUN AND WIND

Sunlight: Orient greenhouses, solar panels, and gardens to maximize exposure. In the northern hemisphere, south-facing structures get the most sunlight.

Example: A solar-powered greenhouse with thermal mass (like concrete or water barrels) can regulate temperatures in cold climates.

Windbreaks: Plant dense trees like spruce or install fences to block prevailing winds. These barriers reduce heating costs by up to 25% in winter.

MULTI-FUNCTIONAL STRUCTURES

Combine uses to save space and resources like a barn with a rainwater collection system that feeds into livestock troughs or a root cellar built beneath a shed for dual-purpose storage.

WATER MANAGEMENT INTEGRATION

Position storage tanks uphill to allow gravity-fed irrigation.

Build a series of ponds at different elevations. Use the overflow from one pond to irrigate fields below.

A well-thought-out homestead design doesn't just save you time and energy—it creates a sanctuary that aligns with nature's rhythms. By embracing advanced zoning, permaculture principles, and efficient land use, your homestead becomes more than a home—it becomes a legacy.

CHAPTER THREE: EXPANDING YOUR GARDEN

Now it's time to elevate your garden into an advanced, self-sustaining powerhouse of productivity. This chapter will guide you through creating systems that require less maintenance, produce more, and work harmoniously with nature. These techniques are perfect for experienced homesteaders ready to grow smarter—not just harder.

Perennial Plants, Trees, and Shrubs

As we touched on in the material about Zones, food forests mimic nature's most productive ecosystems, providing food, habitat, and soil enrichment with minimal input over time. They are built on layers, each contributing to the whole. Here are the layers again for you to review.

LAYERS OF A FOOD FOREST

Canopy Layer: Large fruit or nut trees like walnuts, pecans, or standard-sized apple trees. These provide structure and shade, which can protect understory plants from harsh sun.

Understory Layer: Medium-sized trees like pawpaws or citrus in warm climates. These fit neatly beneath the canopy without overcrowding.

Shrub Layer: Berry bushes such as blueberries, blackcurrants, or elderberries. They thrive in partial shade and produce reliably.

Herbaceous Layer: Think perennial vegetables and herbs, such as asparagus, fennel, and mint. These add diversity and repel pests.

Ground Cover: Cover the soil with plants like creeping thyme, clover, or creeping raspberry. These retain moisture and outcompete weeds.

Root Layer: Add value underground with crops like sunchokes or chicory, which also aerate the soil.

Climbing Layer: Vertical space is valuable real estate! Try hardy kiwi, grapevines, or hops.

Food Forest Plans For Three Climate Zones

A food forest mimics a natural forest's structure and biodiversity while focusing on edible plants. Each layer—from canopy to ground cover—is carefully chosen based on the climate, ensuring resilience and productivity.

NORTHERN COLD CLIMATE FOOD FOREST PLAN

This plan is designed for areas with short growing seasons, cold winters, and moderate rainfall. The focus is on hardy, cold-tolerant plants and species that store well through winter.

Layers and Plant Suggestions

Canopy Layer

Trees: Apple (cold-hardy varieties like Honeycrisp), Pear (Bartlett or Seckel), and Walnut.

Spacing: 15–20 feet apart to maximize sunlight in short growing seasons.

Tip: Plant on the south-facing slope for warmth and extended growing days.

Understory Layer

Small Trees/Shrubs: Hazelnut, Elderberry, and Serviceberry.

Function: Provide nitrogen fixation and bird-attracting berries for pest control.

Shrub Layer

Perennials: Red and black currants, gooseberries.

Function: Excellent for making preserves and high in vitamin C.

Herbaceous Layer

Herbs: Mint, chives, and comfrey.

Tip: Comfrey acts as a dynamic accumulator, pulling nutrients from deep in the soil.

Ground Cover Layer

Crops: White clover (fixes nitrogen) and creeping thyme (suppress weeds and attract pollinators).

Root Layer

Edibles: Jerusalem artichokes, horseradish, and garlic.

Tip: Plant garlic near fruit trees to deter pests.

Climbers/Vines

Plants: Hardy kiwi and grapevines.

Support: Train up the canopy trees or trellises.

Water Management

Swales: Contour swales capture rainwater during spring melts

Mulching: Use straw or wood chips to insulate roots in winter and retain moisture in summer.

Key Strategies

Incorporate windbreaks with dense trees like spruce or pine on the north side.

Use compost piles at the edge of the forest to introduce fertility naturally.

TROPICAL CLIMATE FOOD FOREST PLAN

Tropical food forests thrive on biodiversity and fast-growing plants. The focus here is on year-round productivity and perennial crops that handle high heat and humidity.

Layers and Plant Suggestions

Canopy Layer

Trees: Mango, Breadfruit, Jackfruit, and Coconut. Spacing: Wider spacing (20–30 feet) to allow light for lower layers.

Tip: Stagger planting times for continuous yields.

Understory Layer

Small Trees/Shrubs: Papaya, Guava, and Moringa. Function: Moringa provides protein-rich leaves and acts as a living mulch when pruned.

Shrub Layer

Perennials: Banana, Pineapple, and Coffee plants.

Function: These thrive in partial shade and benefit from the high humidity of tropical forests.

Herbaceous Layer

Herbs: Lemongrass, turmeric, and ginger.

Tip: These plants thrive in damp, shaded conditions and enrich the soil.

Ground Cover Layer

Crops: Sweet potatoes, peanuts, and perennial beans.

Function: Stabilize soil and reduce erosion during heavy rains.

Root Layer

Edibles: Taro, cassava, and yams.

Tip: Root crops thrive in the tropical climate and

offer calorie-dense food.

Climbers/Vines
Plants: Passionfruit and Malabar spinach.
Support: Train up papaya trees or trellises for shade-tolerant climbers.

Water Management

Rain Gardens: Plant near low-lying areas to collect water during downpours.
Irrigation: Drip irrigation systems from rain barrels ensure consistent moisture.

Key Strategies

Incorporate nitrogen-fixing plants like Leucaena (aka "tree lucerne") to replenish soil nutrients.
Intercrop with fast-growing species like corn or pigeon peas for shade and quick yields.

Arid Desert Food Forest Plan

An arid food forest focuses on drought-tolerant species and water-efficient systems to ensure sustainability in low-rainfall regions.

LAYERS AND PLANT SUGGESTIONS

Canopy Layer
Trees: Mesquite, Acacia, Date Palm, and Jujube.
Spacing: Cluster plantings to reduce water evaporation.
Tip: Use shade from taller trees to cool understory plants.

Understory Layer
Small Trees/Shrubs: Pomegranate, Fig, and

Desert Willow.

Function: Provide fruits and habitat for pollinators.

Shrub Layer

Perennials: Prickly pear cactus (nopales), rosemary, and sage.

Function: Provide edible pads and fruit, while rosemary and sage deter pests.

Herbaceous Layer

Herbs: Lavender, oregano, and thyme.

Tip: These thrive in dry soil and provide medicinal and culinary uses.

Ground Cover Layer

Crops: Creeping ground covers like purslane and perennial peanut.

Function: These reduce soil erosion and act as living mulch.

Root Layer

Edibles: Carrots, beets, and desert-adapted potatoes.

Tip: Deep mulching retains moisture for root crops.

Climbers/Vines

Plants: Grapevines (drought-resistant varieties).
Support: Train up Acacia or trellises.

Water Management

Keyhole Gardens: Funnel organic waste and water into central composting areas.

Hügelkultur Beds: Bury wood logs under soil to act as long-term water reservoirs.

Key Strategies

Use greywater systems to recycle household water for irrigation.

Surround the food forest with windbreaks, such as living fences of drought-tolerant shrubs.

Each food forest plan highlights the unique challenges and opportunities of its climate zone. With careful planning and tailored strategies, you can create a thriving, resilient ecosystem that provides food, reduces labor, and regenerates the land.

These systems showcase the power of working with nature, not against it, to create abundance in any climate.

Cover Crops, Green Manures, and Mulching

Healthy soil is the cornerstone of a thriving homestead, and these advanced techniques ensure your soil remains rich, productive, and alive.

Cover crops aren't just for fallow fields—they're essential for keeping your soil healthy.

Best Picks: Crimson clover for nitrogen fixation, rye for erosion control, and buckwheat for fast biomass production.

How-To: Plant a fast-growing cover crop like buckwheat between spring and fall harvests. Mow it down before flowering, and till it in as green manure.

GREEN MANURES AS SOIL BOOSTERS

Green manure crops are your secret weapon for adding organic matter.

Example: Plant hairy vetch in late summer. By spring, it'll be ready to till under, leaving a nitrogen-rich bed for your vegetables.

ADVANCED MULCHING TECHNIQUES

Lasagna Gardening: Layer materials like cardboard, kitchen scraps, straw, and compost directly on your garden beds. Over time, this creates a nutrient-dense planting area.

Wood Chip Mulch: Spread 3–4 inches of untreated wood chips around perennial plants. As they decompose, they mimic forest floor dynamics.

Suggestion: Soil tests are essential for fine-tuning your efforts. Use kits from your local extension office to identify deficiencies.

Greenhouses, Cold Frames, and Hotbeds

Imagine harvesting fresh greens in December or starting your tomatoes months before your neighbors. Season extension tools make this possible, no matter your climate.

GREENHOUSES

DIY Options: Build a hoop house with PVC pipes and plastic sheeting for a cost-effective solution.

Heating Tips: Paint barrels black and fill them with water to absorb and release heat.

Advanced Trick: Use shade cloths in summer to avoid overheating.

COLD FRAMES

Cold frames are perfect for winter greens and extending the fall growing season.

Material Ideas: Repurpose old windows, clear acrylic sheets, or even shower doors.

Crop Suggestions: Try spinach, kale, or mache,

which thrive in cooler temperatures.

HOTBEDS FOR EARLY STARTS

How-To: Dig a trench, fill it with fresh manure, cover with soil, and plant cold-tolerant crops like lettuce or radishes. The decomposing manure provides bottom heat, accelerating growth.

Tip: Use row covers or frost blankets inside cold frames or greenhouses to add extra protection during severe frosts.

Animals for Natural Fertilization

Animals can turn your garden into a closed-loop system, reducing your reliance on store-bought fertilizers while managing pests and weeds.

CHICKENS

Use a chicken tractor to confine chickens to specific areas. They'll scratch up weeds, eat pests, and fertilize the soil.

Tip: Sprinkle seeds of cover crops in the area before moving the tractor. The chickens will scratch them in for you.

DUCKS

Ducks love slugs and other pests but are gentler on plants than chickens. They're perfect for orchards and berry patches.

Breed to Try: Indian Runner Ducks are prolific pest controllers.

RABBITS

Their droppings can be applied directly to the garden as a rich, balanced fertilizer.

Housing Idea: Keep them in mobile hutches that can be moved over garden beds.

GOATS

Use goats for clearing invasive plants. Their manure adds valuable nutrients to compost piles.

Caution: Goats can be destructive—use temporary fencing to control their grazing.

By layering advanced techniques like food forests, soil management, season extension, and animal integration, your garden can evolve into a regenerative system that thrives year after year. These strategies aren't just about productivity—they're about building a sustainable future.

Every new technique you implement is an opportunity to learn, adapt, and create a garden that reflects your ingenuity and hard work.

CHAPTER FOUR: ANIMAL HUSBANDRY

Advanced animal husbandry builds on foundational knowledge, focusing on larger livestock, dairy production, animal health, and sustainable breeding. This chapter delves deeply into these topics to help you elevate your homestead to new levels of productivity, sustainability, and self-sufficiency.

Raising Goats, Sheep, and Pigs

Transitioning to larger livestock introduces exciting opportunities for your homestead but requires careful planning and resource management. These animals bring diverse benefits, from milk and meat to wool and land improvement.

GOATS

Goats are the Swiss army knife of livestock—versatile, hardy, and rewarding. Their milk, meat, and weed-control abilities make them a favorite among homesteaders.

Some reasons to incorporate goats into your homestead are:

Milk Production: Goats produce rich, creamy milk ideal for drinking, cheesemaking, and yogurt.

Meat: Goat meat (chevon) is lean, flavorful, and sought after in many cuisines.

Land Clearing: Goats are natural browsers, ideal for clearing brush, weeds, and invasive plants.

Housing and Fencing

Space Requirements: At least 250 square feet per goat for outdoor browsing.

Fencing Tips:

Minimum height: 4 feet. Add electric wire at the

top to deter climbers.

Use woven wire fencing with no larger than 4-inch openings.

Regularly inspect and repair fences; goats test boundaries constantly.

Feed and Nutrition

Browse over Graze: Goats prefer shrubs, weeds, and leaves. Supplement with hay in winter.

Mineral Access: Provide free-choice goat-specific mineral blocks to prevent deficiencies (e.g., selenium or copper).

Breeds for Specific Goals

Dairy: Nubian (high butterfat) and Saanen (high milk yield).

Meat: Boer goats are robust and fast-growing.

Dual-Purpose: Kinder goats offer moderate milk and meat production.

SHEEP

Sheep bring a trifecta of benefits to the homestead: high-quality meat, milk for specialty cheeses, and wool for crafting. Their gentle nature and efficient grazing make them ideal for small or large properties.

Here are some benefits to bringing sheep onto your homestead:

Efficient Grazers: Perfect for maintaining pastures.

Versatile Production: One animal can provide food, fiber, and milk.

Relatively Low Maintenance: Require less intensive care compared to goats or pigs.

Housing and Space Requirements

Minimum 20 square feet per sheep indoors, plus

ample outdoor pasture.

Provide shelters that protect from rain, wind, and predators.

Breed Selection

Meat: Dorper and Katahdin are known for rapid growth and tender lamb.

Dairy: East Friesian sheep are prolific milkers.

Wool: Romney and Icelandic breeds offer high-quality fleece.

Shearing and Wool Care

Shear once annually, ideally in spring before lambing.

Electric clippers streamline the process; practice ensures efficiency.

Sell raw fleece to fiber enthusiasts or process it for personal use.

PIGS

Pigs are invaluable for recycling food waste, enriching the soil, and providing premium meat. Their intelligence and adaptability make them both a challenge and an asset.

Why consider pigs for your homestead?

Meat Production: Pigs grow quickly and provide abundant pork.

Soil Enrichment: Their rooting behavior tills soil and incorporates organic matter.

Waste Management: Pigs can consume food scraps and garden waste, reducing landfill contributions.

Housing and Fencing

Secure housing with dry bedding is essential; pigs are sensitive to temperature extremes.

Use electric fencing to confine pigs safely and

humanely.

Feeding Strategies

Supplement commercial pig feed with kitchen scraps, garden surplus, and whey from cheesemaking.

Avoid feeding raw potatoes, chocolate, or avocado, which can harm pigs.

Recommended Breeds

Meat: Berkshire and Tamworth are prized for flavor.

Foraging: Kunekune pigs thrive on pasture, requiring less supplemental feed.

Dairy Production on the Homestead: Milking, Cheesemaking, and Yogurt

Adding dairy production to your homestead enriches your diet and opens opportunities for artisanal products. Milking goats or sheep and creating homemade dairy products like cheese and yogurt are gratifying and cost-effective.

MASTERING MILKING TECHNIQUES

Milking requires consistency, cleanliness, and patience.

Essentials

Milking Stanchion: A sturdy stanchion ensures the animal stays calm and still.

Routine: Milk at the same time daily to maintain consistent production.

Cleaning: Wash teats with warm water and a mild antiseptic solution.

Milking Methods

Hand-Milking: Ideal for beginners or small herds.

Milking Machines: Save time for larger operations; ensure proper cleaning to prevent contamination.

CHEESEMAKING AT HOME

Homemade cheese adds value to surplus milk and introduces artisanal flavors to your homestead kitchen.

Essential Tools

Cheesemaking pot (stainless steel), thermometer, cheesecloth, and starter cultures.

Beginner-Friendly Cheeses
Ricotta
Chèvre:

YOGURT

Transform milk into creamy yogurt with these simple steps:

Heat milk to 180°F, cool to 110°F.

Stir in plain yogurt with live cultures or a powdered starter.

Incubate in a warm place (e.g., yogurt maker or oven with light on).

Animal Health

Keeping animals healthy is the cornerstone of a productive and thriving homestead. Advanced health care for homestead animals blends time-tested natural remedies, preventative measures like vaccination, and professional veterinary care to ensure the wellbeing of your livestock.

NATURAL REMEDIES

Natural remedies are an essential part of homestead animal care, offering eco-friendly and cost-effective solutions. When used correctly, they can boost immunity, treat minor ailments, and improve overall health.

Herbal Remedies

Garlic: Acts as a natural dewormer and immune booster. Add crushed garlic to feed in small amounts for chickens, goats, and sheep.

Echinacea: Stimulates the immune system during cold and flu seasons. Create a tea from dried echinacea flowers and mix it into drinking water.

Calendula: Use calendula salve to treat minor cuts, scrapes, and skin irritations in all types of livestock.

Essential Oils

Tea Tree Oil: Treats fungal infections and repels parasites when diluted and applied topically.

Lavender Oil: Reduces stress in animals and can be used in nesting areas to calm poultry.

Peppermint Oil: Helps with respiratory issues; use sparingly in goats and sheep.

Probiotics and Fermented Foods

Adding probiotics to feed improves gut health, especially in poultry and ruminants. Fermented feed (e.g., lacto-fermented grains for chickens) can boost nutrient absorption and prevent digestive issues.

Mineral Supplements

Diatomaceous Earth: Acts as a natural wormer when mixed into feed at recommended doses.

Seaweed or Kelp Meal: Provides essential trace

minerals, improving coat condition and fertility in larger livestock like cattle and horses.

HOME REMEDIES FOR COMMON AILMENTS

Scours in Calves/Goats: Treat with a mixture of activated charcoal, yogurt, and electrolyte solutions.

Mastitis: Apply warm compresses of calendula tea and massage with olive oil infused with rosemary.

Fly Control: Create a natural fly spray using apple cider vinegar, essential oils (lavender or citronella), and water.

VACCINATION

Vaccination is a crucial preventative measure, especially for livestock prone to contagious or fatal diseases. While it may seem unnecessary for small-scale homesteads, the benefits far outweigh the risks, particularly if your animals are exposed to wildlife or neighboring farms.

Essential Vaccines

Cattle: Clostridial vaccines (e.g., Blackleg), respiratory vaccines (IBR, BVD).

Sheep and Goats: CD&T vaccine (Clostridium perfringens types C & D and tetanus).

Chickens: Marek's disease vaccine and Newcastle disease vaccine.

Pigs: Erysipelas and porcine parvovirus vaccines.

Vaccination Tips

Timing: Follow species-specific schedules. For example, vaccinate lambs and kids between 4–6 weeks and boost annually.

Record Keeping: Maintain detailed records of each animal's vaccinations, including batch numbers and dates.

Safe Storage: Store vaccines in a temperature-controlled refrigerator to ensure efficacy.

NATURAL VACCINATION ALTERNATIVES

Though limited, natural immunization techniques, such as controlled exposure to mild pathogens, can sometimes strengthen immunity. For example, chickens allowed to forage develop stronger resistance to coccidiosis.

VETERINARY CARE

No matter how skilled you are with natural remedies, there will be times when professional veterinary care is essential. Building a strong relationship with a trusted livestock vet ensures you're prepared for emergencies and can address complex health issues promptly.

Routine Veterinary Services

Health Checks: Schedule annual wellness exams for large animals to monitor overall health and detect early signs of illness.

Hoof Care: Ensure proper trimming and treatment for hoof rot or overgrowth in goats, sheep, and cattle.

Reproductive Health: Have breeding stock checked for fertility and monitor pregnant animals closely for complications.

Emergency Situations

Signs to Call a Vet: Sudden lameness, prolonged labor, severe wounds, or unresponsiveness are red flags.

Emergency Preparedness Kit: Stock items like sterile gauze, antibiotic ointments, syringes, and electrolyte powders.

Diagnostic Tools

Use portable tools like stethoscopes, thermometers, and fecal egg count kits for on-the-spot assessments.

Work with your vet to interpret results from blood tests or fecal analyses.

Integrated Health Management: Prevention is Key
Combining natural remedies, vaccination, and veterinary care is the best way to ensure your animals remain healthy and productive. Here are strategies for creating a robust health management plan:

HYGIENE AND BIOSECURITY

Keep barns and coops clean to prevent the spread of diseases.

Quarantine new animals for at least 2–4 weeks before integrating them into the herd or flock.

PASTURE MANAGEMENT

Rotate pastures to reduce parasite loads and maintain forage quality. Use integrated livestock systems (e.g., chickens following cattle) for pest control.

OBSERVATION AND RECORD KEEPING

Monitor animals daily for signs of illness or distress. Keep detailed records of feed, medical treatments, and behavior changes.

COMMUNITY SUPPORT

Join local homesteading or livestock groups to share knowledge and get recommendations for reliable veterinarians or suppliers.

A holistic approach to animal health ensures your

livestock are not only disease-free but thriving. By combining natural remedies, essential vaccinations, and proactive veterinary care, you'll create a resilient and productive homestead ecosystem.

The best health management practices start with observation, education, and a willingness to adapt as needed. Healthy animals mean higher yields, less stress, and a more sustainable homestead.

Sustainable Breeding Practices

Sustainable breeding practices are a cornerstone of a thriving and balanced homestead. Beyond maintaining the productivity of your herd, sustainable breeding ensures the health and welfare of your animals, preserves genetic diversity, and aligns with ethical and ecological principles.

SELECTIVE BREEDING GOALS

The foundation of sustainable breeding lies in clear, purposeful goals tailored to your homestead's needs. By focusing on specific traits, you can improve productivity, health, and adaptability while avoiding genetic pitfalls like inbreeding.

Identifying Traits to Improve

Productivity: Prioritize traits like milk production, egg-laying frequency, or meat yield.

Disease Resistance: Select animals that demonstrate resilience to local diseases or pests.

Temperament: Breed animals that are docile and easy to handle, especially for small homesteads.

Climate Adaptability: Choose breeds suited to your environment (e.g., heat-tolerant breeds for arid regions).

Complementary Pairing

Pair animals with complementary traits to enhance overall performance.

Example: Breed a high-yield dairy goat (e.g., Nubian) with a hardy meat breed (e.g., Boer) to produce dual-purpose offspring.

Avoid pairing animals with overlapping genetic weaknesses, such as susceptibility to parasites or poor hoof health.

Genetic Diversity

Rotate breeding stock or introduce new genetics periodically to prevent inbreeding.

Maintain accurate breeding records to track lineage and avoid accidental inbreeding.

SUSTAINABLE BREEDING STRATEGIES

Optimal Breeding Ages

Breeding animals too young can result in complications, while breeding animals too old can reduce fertility and offspring viability.

Cattle: Breed heifers at 15–18 months old.

Goats and Sheep: Wait until does and ewes are at least 8–12 months old, depending on breed and size.

Poultry: Hens are most fertile between 6 months and 2 years.

Frequency of Breeding

Overbreeding depletes the health of your animals. Use seasonal breeding to align with natural cycles and pasture availability.

Allow a rest period between pregnancies, particularly for high-yielding animals like dairy goats and cows.

Culling for Sustainability

Regularly assess the productivity and health of breeding stock.

Remove animals with chronic health issues, poor fertility, or subpar offspring from the breeding program.

Managing the Breeding Process

Natural Breeding: Preferred for most homesteads due to its simplicity and lower cost.

Artificial Insemination (AI): Allows access to superior genetics without purchasing new animals. Ideal for dairy cattle or rare breeds.

Controlled Breeding: Use separate pens or paddocks to prevent unplanned matings.

Timing Breeding Cycles

Synchronize breeding to align births with optimal forage availability or weather conditions.

Use visual or behavioral cues (e.g., heat cycles in goats or sheep) to determine the best time to mate.

MANAGING BIRTHING

Proper birthing management ensures a safe and healthy delivery for both mother and offspring. Preparation, monitoring, and post-birth care are critical.

Preparation

Set up a clean, quiet birthing area equipped with soft bedding, fresh water, and minimal disturbance.

Stock a birthing kit with essentials: sterile gloves, iodine for navel dipping, clean towels, and a bulb syringe for clearing airways.

Signs of Labor
Watch for physical and behavioral changes:
Swollen udder or vulva.
Restlessness, pacing, or nesting behavior.
Discharge or visible contractions.
Assisting During Labor

Most animals deliver without assistance, but intervene if labor lasts unusually long (e.g., more than 30 minutes of active pushing).

Learn basic techniques for repositioning a stuck kid, lamb, or calf in cases of dystocia (difficult birth).

Post-Birth Care

Ensure the newborn begins nursing within the first hour to receive colostrum, which is vital for immunity.

Monitor the mother for signs of retained placenta or excessive bleeding.

RAISING HEALTHY OFFSPRING

Nutrition

Provide high-quality feed and supplements to nursing mothers to ensure sufficient milk production.

Transition weaned animals to appropriate forage and grains gradually to avoid digestive issues.

Early Health Interventions

Vaccinate offspring according to species-specific schedules.

Castrate males not intended for breeding early to minimize stress and avoid behavioral issues.

Socialization

Socialize young animals with both humans and herd mates to reduce stress and promote bonding.

ETHICAL CONSIDERATIONS IN SUSTAINABLE BREEDING

Avoiding Overpopulation

Only breed animals when you have a clear plan for their use (e.g., milk production, meat, or sale.

Find responsible buyers for surplus animals, ensuring they'll receive proper care.

IMPROVING ANIMAL WELFARE

Avoid breeding animals with known genetic disorders or traits that compromise their health.

Maintain humane practices during breeding, birthing, and culling.

BALANCING PRODUCTIVITY AND SUSTAINABILITY

Don't prioritize production traits (e.g., maximum milk yield) at the expense of long-term health and hardiness.

Ensure breeding decisions support the overall ecological balance of your homestead.

Sustainable breeding is an art and science that requires thoughtful planning, observation, and care. By focusing on complementary traits, maintaining genetic diversity, and prioritizing the health of your animals, you'll create a thriving herd that serves your homestead for years to come.

With responsible breeding practices, you're not just building a herd—you're cultivating a legacy of productivity, resilience, and sustainability.

CHAPTER FIVE: ADVANCED BEEKEEPING

Beekeeping isn't just a hobby; it's an advanced skill that bridges sustainable living, agriculture, and ecological preservation.

This chapter moves beyond basic hive care, diving into nuanced methods to supercharge honey production, fortify your hives against challenges, and create a thriving pollination ecosystem for your homestead and surrounding areas.

Mastering Hive Management for Peak Productivity

SWARM CONTROL

Swarming is one of the biggest threats to productivity, but advanced strategies can help mitigate it while maintaining colony strength.

REQUEENING PRACTICES

Every 1–2 years, replace queens to prevent the colony from naturally initiating swarming. Use a young queen from a strong lineage to maintain vitality. Consider raising your own queens if you're managing multiple hives—it's cost-effective and allows control over genetics.

The Miller Method

The Miller Method for queen rearing is a straightforward and effective technique designed for small-scale beekeepers who want to raise their own queens without needing specialized equipment or advanced skills. Named after Dr. C.C. Miller, a prominent beekeeper and author, this method utilizes natural bee behaviors to rear queens, making it both beginner-friendly and scalable for advanced practices. This technique is particularly useful for beekeepers involved in requeening, expanding their apiary, or maintaining genetic diversity in their hives.

The Miller Method involves creating queen cells by using a comb with staggered, natural edges, which the bees can easily transform into queen cells. It mimics the bees' natural tendency to rear new queens in response to the absence of a queen or the presence of young larvae in the colony.

Unlike other queen-rearing methods, such as grafting, the Miller Method eliminates the need for tools like grafting needles or cell cups. Instead, it relies on the bees' instincts, allowing them to choose and nurture larvae into queens with minimal intervention.

HOW THE MILLER METHOD WORKS

Preparing the Comb

Start with a clean frame fitted with wax foundation or drawn comb. Cut the comb into triangular or staggered edges. These irregular edges provide the bees with natural starting points for queen cell construction.

Selecting the Larvae

Choose a donor colony with desirable traits (e.g., gentle temperament, disease resistance, or high honey production). The comb should contain young larvae less than 24 hours old, as these are most likely to be successfully raised into queens.

Installing the Prepared Comb

Place the prepared frame into a queenless colony or a designated queen-rearing starter colony. Ensure the colony is strong, with a good population of nurse bees and abundant resources (nectar and pollen).

Bees Start Queen Rearing

The absence of a queen triggers the colony's natural response to rear a new one. The nurse bees will select the most suitable larvae from the comb and begin to enlarge the cells, feeding the larvae royal jelly to develop them into queens.

Monitoring Queen Cell Development

Over the next 5–10 days, the larvae will develop into queen pupae. Check the colony to monitor the progress of the queen cells. Ensure that the colony remains strong and well-fed during this period.

Harvesting Queen Cells

Once the queen cells are capped (around day 8–9), they can be transferred to mating nucs or other colonies for hatching. Handle the queen cells gently to avoid damaging them.

Introducing New Queens

After the queens emerge, they can be introduced to colonies in need of requeening or used to establish new colonies.

BENEFITS OF THE MILLER METHOD

Simplicity

The Miller Method is accessible to beekeepers at all experience levels. It requires no special tools or advanced skills, relying instead on the natural behaviors of bees.

Cost-Effectiveness

Unlike methods that require grafting tools, cell cups, or incubators, the Miller Method uses existing hive components and resources.

Natural Queen Rearing

This method allows bees to choose the best larvae for queen rearing, ensuring healthy and robust queens.

Flexibility

Suitable for small-scale operations, the method can be scaled up for larger apiaries by using multiple queen-rearing colonies.

Improved Hive Genetics

Beekeepers can selectively breed queens with desirable traits, enhancing the overall health and productivity of their apiary.

PRACTICAL TIPS FOR SUCCESS

Timing Is Critical

Perform the Miller Method during the peak brood-rearing season, typically in spring or early summer when colonies are strong, and resources are abundant.

Ensure Colony Strength

Use a strong, queenless starter colony with plenty of nurse bees to rear healthy queen cells. A weak colony may struggle to produce quality queens.

Maintain Optimal Conditions

Provide ample nectar and pollen sources to support the intensive feeding of larvae with royal jelly. Consider supplementing with sugar syrup if natural forage is limited.

Control Swarming

Ensure the starter colony does not become overcrowded, as this could lead to swarming behavior.

Protect Queen Cells

Queen cells are delicate, especially during capping and pupation. Handle them with care and avoid exposing them to extreme temperatures.

Monitor Queen Emergence

Once queens emerge, they may fight to establish dominance. If raising multiple queens, transfer queen cells or virgin queens to separate mating nucs to prevent losses.

CHALLENGES AND HOW TO OVERCOME THEM

Rejected Queen Cells

Occasionally, bees may reject or destroy some queen cells. Ensure the colony is truly queenless and strong enough to rear queens.

Poor-Quality Queens

The health of the donor colony and availability of resources directly impact queen quality. Choose donor colonies with robust genetics and ensure abundant resources in the rearing colony.

Timing Errors

Missing the window for harvesting queen cells can lead to virgin queens emerging and fighting. Mark your calendar and monitor progress closely.

The Miller Method offers an elegant, natural approach to queen rearing that aligns with the instincts of honeybees. Its simplicity, cost-effectiveness, and reliability make it an excellent choice for small-scale and advanced beekeepers alike. By mastering this method, you can maintain strong colonies, improve hive genetics, and ensure the sustainability of your apiary.

The Demaree Method

Now, let's dive into one of my favorite brood management methods in my own hives: employing the Demaree Method for swarm prevention.

The Demaree Method was developed by George Demaree, an American beekeeper and innovator, in 1892. Demaree designed this method to address one of the most common challenges faced by beekeepers: preventing swarms without compromising the colony's productivity.

This technique centers around separating the queen from the majority of the brood, which calms the bees' swarming instinct while maintaining the colony as a single, cohesive unit. The method's simplicity and effectiveness have made it a staple for modern beekeepers aiming to optimize hive management during the peak season.

George Demaree was a pioneer in practical beekeeping during the late 19th century. At the time, swarming was considered an inevitable and often frustrating aspect of beekeeping. Demaree recognized the need for a method to keep colonies intact while alleviating the conditions that prompt swarming. His innovative approach, which involves rearranging brood and confining the queen, marked a significant advancement in hive management and swarm control.

HOW THE DEMAREE METHOD WORKS

Identify the Queen

Begin by locating the queen. Accurate identification is essential for implementing the method successfully.

Reorganize the Hive

Move the queen and several frames of open brood

(eggs and larvae) to a lower brood box. This box is placed on the bottom board. Place the remaining capped brood, honey, pollen, and empty comb in an upper brood box.

Install a Queen Excluder
Insert a queen excluder between the lower box (with the queen) and the upper box. This keeps the queen confined to the lower box while allowing worker bees to move freely between the two boxes.

Add Honey Supers (Optional)
If honey production is a goal, place honey supers above the upper brood box to provide space for nectar storage.

Inspect Regularly
Check the hive weekly to ensure no queen cells are being raised in the upper brood box. Remove any queen cells to prevent swarming.

The Demaree Method provides an ideal setup for raising queens. By separating the queen from most of the brood, the upper box becomes a space where nurse bees can focus on raising queen cells. These cells can be harvested and used to requeen hives or start new colonies. It also helps stabilize the colony, creating the strong workforce needed for successful queen rearing using the Miller Method.

The division of brood in the Demaree Method ensures a steady supply of worker bees, which can support both honey production and queen-rearing projects.

BENEFITS OF THE DEMAREE METHOD

Swarm Prevention
Swarming reduces a colony's workforce,

impacting productivity. The Demaree Method alleviates overcrowding, a primary trigger for swarming.

Enhanced Hive Productivity
Keeping the colony together ensures a robust foraging force and efficient hive operations, leading to higher honey yields.

Flexible Management
The method allows for strategic reorganization of hive resources, enabling beekeepers to focus on honey production, brood development, or queen rearing as needed.

No Splits Required
Unlike other swarm control techniques, the Demaree Method avoids splitting the colony, making it ideal for beekeepers who want to maintain hive strength.

Natural Colony Dynamics
By preserving the original colony structure, this method minimizes disruption to the bees' natural behavior.

CHALLENGES OF THE DEMAREE METHOD

Labor-Intensive
Weekly inspections are essential to ensure queen cells are not being raised in the upper box.

Potential for Missed Queen Cells
If queen cells are overlooked, the colony may still swarm.

Bee Stress
Rearranging the hive can temporarily disrupt

colony dynamics, especially during a nectar dearth or adverse weather.

Requires Careful Timing
The method is most effective when implemented before the colony becomes overcrowded or exhibits strong swarming impulses.

Imagine a beekeeper noticing early signs of swarming in a strong hive. They implement the Demaree Method to separate the queen and open brood, reducing the colony's swarming urge. Simultaneously, they use the Miller Method by introducing a comb with young larvae into the upper brood box. This encourages the bees to raise new queens while the hive remains productive and intact.

The beekeeper can later harvest the queen cells for requeening or starting nucleus colonies, turning a potential swarm into an opportunity for hive expansion.

The Demaree Method is a versatile and effective tool for managing swarming and brood distribution in a hive. Its principles complement queen-rearing techniques like the Miller Method, creating a dynamic strategy for advanced beekeeping. By mastering these methods, beekeepers can achieve a balance between preventing swarms, raising high-quality queens, and maintaining robust, productive colonies.

PEST AND DISEASE PROTOCOLS

Pests like varroa mites and diseases such as chalkbrood require constant vigilance.

Mite-Resistant Bees: Breeds like the VSH (Varroa Sensitive Hygiene) or Minnesota Hygienic bees are naturally adept at cleaning out infected cells, keeping mite levels low.

Integrated Pest Management (IPM)

Rotate treatment types (chemical, biological, and mechanical) to prevent pests from developing resistance.

Incorporate screened bottom boards in your hives, which allow mites to fall out and prevent reinfestation.

Use drone comb removal to interrupt the mite breeding cycle, as mites prefer drone brood cells.

ENHANCING COLONY STRENGTH

Building strong colonies is a cornerstone of advanced beekeeping.

Feeding Programs for Brood Growth: Instead of plain sugar syrup, enrich feeding solutions with pollen patties or protein supplements during early spring.

DIY Recipe for Pollen Patties: Mix powdered sugar, pollen substitute, and a small amount of honey to form pliable patties. Place them directly over the brood nest.

Hive Manipulation: Perform "checkerboarding," where empty frames alternate with brood frames in the spring. This prevents overcrowding and stimulates expansion.

Maximizing Pollination Impact

SUPERCHARGING CROP POLLINATION

Crop Pairing: Optimize pollination by planting crops in proximity to attract bees:

Position high-nectar plants like buckwheat near

orchard trees for overlapping bloom cycles.

For vegetable gardens, intersperse flowering herbs like basil and thyme to encourage frequent visits.

Selective Forage Management: Not all flowers provide equal benefits. Focus on plants with high pollen and nectar content.

Example Plants: Clover, alfalfa, and borage are excellent bee forage plants that bloom over long periods.

DESIGNING A POLLINATOR PARADISE

Establish Permanent Foraging Areas: Allocate 10–15% of your homestead space to wildflower meadows or hedgerows.

Tip: Incorporate a mix of annuals and perennials to ensure year-round forage. Wildflowers like black-eyed Susan, yarrow, and purple coneflower support bees through droughts and dearths.

Water Sources: Bees need water for cooling the hive and diluting honey. Install shallow water stations with floating corks or small rocks for safe landings.

COLLABORATING WITH NEIGHBORS

Work with local farmers or gardeners to expand pollination zones. If they grow pesticide-heavy crops, educate them on bee-friendly practices like spraying during early morning hours when bees are least active.

Honey Harvesting and Processing

OPTIMAL HARVEST TIMING

Monitoring Moisture Levels: Use a refractometer to check honey moisture before harvesting. Honey with less than 18% moisture resists fermentation.

Flow Hive Systems: Advanced setups like Flow Hives allow for easy honey extraction without opening the hive. These systems reduce stress on the colony and save time.

REFINING EXTRACTION TECHNIQUES

Radial vs. Tangential Extractors: For larger operations, radial extractors allow you to extract honey from multiple frames at once while preserving comb integrity.

Advanced Wax Processing: After extraction, purify wax by melting it in a solar wax melter. Reuse it for making candles, balms, or starter sheets for new hives.

HONEY VARIATIONS AND PRODUCTS

Creamed Honey: Use controlled crystallization to create spreadable honey. Seed your batch with fine-grain honey and stir regularly.

Infused Honey: Experiment with infusing honey with flavors like lavender, cinnamon, or vanilla.

Specialized Bee Breeds and Advanced Genetics

ITALIAN VS. CARNIOLAN BEES

Italians are prolific honey producers but consume more winter stores. Carniolans, in contrast, conserve resources and adapt well to fluctuating nectar flows.

Tip: Consider hybrid colonies that combine traits, such as the docility of Italians with the resilience of Carniolans.

RAISING YOUR OWN QUEENS

Establish a small queen-rearing operation:
Use a "queen castle" hive, divided into

compartments, to raise multiple queens simultaneously.

Graft larvae from your strongest colonies into queen cups and place them in a cell-building hive.

Year-Round Hive Management

WINTERIZING HIVES

Moisture Control: Install moisture boards or quilt boxes filled with wood shavings to absorb excess humidity.

Cluster Management: Reduce entrance size to prevent cold drafts and block out predators like mice.

SPRING RECOVERY TACTICS

Inspect for signs of winter die-off:

If hives are weak, combine them with stronger colonies using the newspaper method (a single sheet of newspaper between hive boxes allows slow integration).

Advanced beekeeping is a delicate balance of science, observation, and hands-on care. With these detailed strategies, you'll transform your apiary into a thriving hub of productivity, pollination, and ecological harmony. As you implement these techniques, remember that every adjustment you make not only enhances your hive's success but also contributes to the larger ecosystem.

CHAPTER SIX: RENEWABLE ENERGY

Renewable energy is the backbone of true self-sufficiency, offering independence from utility companies and a cleaner, more sustainable way to power your life. In this chapter, we'll dive into advanced techniques to harness solar, wind, and hydro energy, manage off-grid systems effectively, and calculate your energy needs.

Whether you want to reduce your carbon footprint or prepare for grid outages, mastering these skills will illuminate your homestead—literally and figuratively.

Solar Power

Solar energy is often the first step for homesteaders venturing into renewable energy. But now, we're elevating from basic setups to more advanced, efficient systems.

CHOOSING THE RIGHT SOLAR PANELS

Choosing the right solar panels isn't just about cost or brand; it's about optimizing your specific location's energy potential.

Monocrystalline Panels: These are the most efficient and space-saving. Perfect if you have limited space but higher energy demands.

Polycrystalline Panels: A bit less efficient but more budget-friendly, ideal for expansive setups.

Thin-Film Panels: Lightweight and flexible, great for unconventional placements or portable needs, though less energy-dense.

Tip: Use tools like PVWatts or Solar.com to calculate the exact energy output potential for your specific area based on sunlight exposure.

SOLAR INSTALLATION

1. Site Assessment

Use tools like the **SolarPathfinder** or apps such as **SolarEdge** to assess shading, peak sunlight hours, and optimal placement.

If your property experiences variable weather, consider tracking mounts that follow the sun throughout the day to maximize efficiency.

Tip: Plant low-height crops or flowers beneath raised panels to optimize dual-use land space

2. Mounting Options

Roof Mounts: Save ground space but require a professional inspection to ensure your roof can handle the load.

Ground Mounts: Easier to access for cleaning and repairs; use adjustable mounts to optimize seasonal angles.

DIY Solar Trackers: Build motorized trackers with stepper motors and light sensors to keep panels aligned with the sun.

3. Wiring & System Design

Connect panels in series for higher voltage to reduce energy loss over long distances. Use parallel connections for higher amperage if running multiple high-demand appliances.

Install an **MPPT** (Maximum Power Point Tracking) charge controller for efficiency. Pair it with pure sine wave inverters for sensitive electronics.

MAINTENANCE MASTERY

Cleaning: Wash panels monthly with distilled water and a non-abrasive cloth to avoid buildup.

Monitoring: Install remote monitoring systems like SolarEdge or Enphase apps to track performance.

Agrivoltaics

Agrivoltaics, also known as agro-photovoltaics (APV), is an innovative approach that combines solar power generation with agricultural production. By strategically situating solar panels on farmland, this dual-use system maximizes land productivity, reduces energy costs, and creates a synergistic relationship between renewable energy and farming.

This system represents a major leap forward in sustainability, addressing two pressing global challenges: the need for renewable energy and the demand for increased food production. Here's a deep dive into what agrivoltaics is, how it works, and its transformative potential on the homestead and beyond.

THE BASICS OF AGRIVOLTAICS

Agrivoltaics involves the co-location of solar panels with crops, grazing animals, or other agricultural activities. Unlike traditional solar farms, which dedicate land exclusively to energy production, agrivoltaic systems allow for simultaneous farming activities beneath and around solar panels.

DESIGNING

Elevated or Adjustable Solar Panels: Panels are installed at a height or angle that permits sunlight to reach crops or grazing areas below.

Optimized Spacing: Rows of panels are spaced to balance solar energy capture with agricultural productivity.

Dual Benefits: Panels provide shade and wind protection for plants and animals, while generating clean energy.

KEY BENEFITS OF AGRIVOLTAICS

Efficient Land Use: Agrivoltaics maximizes the productivity of land, allowing farmers and homesteaders to generate electricity while growing crops or raising livestock. This approach is especially valuable in regions where arable land is limited or expensive.

Improved Microclimate for Crops: Solar panels create partial shade, reducing heat stress on crops during hot days and minimizing water evaporation. Panels can also shield crops from extreme weather, such as hail or heavy rain, improving resilience.

Energy Independence: Agrivoltaic systems allow homesteaders to produce their own electricity, reducing reliance on grid power and lowering energy costs. Excess energy can be sold back to the grid or stored in batteries for later use.

Enhanced Crop Yields in Some Scenarios: Studies show that certain crops, such as lettuce, spinach, and tomatoes, can thrive under the partial shade provided by solar panels. The cooler microclimate can extend the growing season and improve water efficiency.

Carbon Reduction: By integrating solar power, agrivoltaics contributes to renewable energy goals, reducing carbon emissions associated with traditional energy production.

IMPLEMENTING AGRIVOLTAICS ON THE HOMESTEAD

1. Choosing the Right System

Fixed-Tilt Systems: Panels are mounted at a fixed angle, suitable for crops that benefit from partial shade, such as leafy greens.

Tracking Systems: Panels that follow the sun's movement can optimize energy production while maintaining consistent shading. These are ideal for large-scale setups.

Elevated Arrays: Raising panels high enough to accommodate livestock grazing underneath is an excellent option for homesteaders with livestock operations.

2. Selecting Compatible Crops and Livestock

Crops: Shade-tolerant crops like broccoli, kale, and beans perform well under panels. Some fruits, such as berries, also benefit from the cooler microclimate.

Livestock: Sheep are particularly compatible, as they can graze beneath panels without damaging the equipment. Chickens can also forage under arrays, contributing to pest control and fertilization.

3. Designing the Layout

Space solar panels to ensure adequate sunlight for crops. For example, rows of panels spaced 10-12 feet apart allow sufficient light penetration while leaving room for farm equipment to pass through.

In areas prone to drought, consider pairing panels with water catchment systems to channel rainwater into irrigation tanks.

4. Combining Energy Storage and Usage

Install batteries to store excess energy for use during peak demand or at night. Use the generated electricity to power irrigation pumps, greenhouse fans, or electric fencing, making your homestead more self-sufficient.

EXAMPLES OF AGRIVOLTAIC SUCCESS

Arid Climates

In arid regions, agrivoltaics has been used to reduce water evaporation and improve crop resilience. For instance, solar panels shading lettuce fields in Arizona reduced water usage by 15-20% while maintaining high crop yields.

Cold Climates

In colder regions, agrivoltaics can extend the growing season by creating a warmer microclimate beneath the panels. This setup is particularly effective for hardy crops like kale and carrots.

Livestock Integration

Sheep grazing under solar panels in Germany helped manage vegetation while providing the animals with shade, improving their welfare and maintaining clean energy production.

CHALLENGES AND CONSIDERATIONS

Initial Investment: Installing an agrivoltaic system requires a higher upfront cost compared to traditional farming or standalone solar panels. However, tax credits and renewable energy incentives can offset this expense.

Maintenance: Maintaining both the panels and the agricultural operations requires coordination. For example, livestock may occasionally damage equipment, necessitating robust fencing and monitoring.

Crop Compatibility: Not all crops thrive in partial shade. Understanding the needs of your plants is critical to maximizing both energy and agricultural output.

For homesteaders, agrivoltaics represents an exciting opportunity to blend energy independence with sustainable agriculture. By implementing this system, you can expect to generate clean energy to power your homestead, improve the resilience and productivity of your crops or livestock and contribute to a more sustainable future.

With careful planning and execution, agrivoltaics can transform your homestead into a model of innovation, efficiency, and environmental stewardship. Whether you're farming leafy greens, raising sheep, or cultivating berries, this dual-use system offers immense potential for those committed to advanced homesteading practices.

Wind Power and Micro-Hydro Solutions

When the wind blows or water flows, you can harness it! Diversifying with wind or hydro power enhances reliability and reduces dependence on any single system.

WIND POWER

Wind energy is particularly effective in regions with consistent speeds above 10–12 mph.

Types of Wind Turbines

Horizontal-Axis Turbines: High efficiency but require consistent, strong winds.

Vertical-Axis Turbines: Compact, quiet, and better suited for turbulent or variable winds.

Tip: Check Wind Maps or install an anemometer

to measure wind speeds over a few months before investing.

INSTALLATION

Turbine Placement: Mount turbines at least 30 feet above obstructions for optimal airflow.

Foundation Setup: Pour a concrete base and secure with guy wires to withstand storms.

Integration: Hybrid systems (wind + solar) ensure consistent power regardless of weather conditions.

Tip: Brands like Bergey and Primus offer homestead-friendly turbine models.

Micro-Hydro Solutions

Micro-hydro is a game-changer for homesteaders lucky enough to have a flowing water source. It provides consistent, weather-independent energy.

WATER SOURCE ASSESSMENT

Measure head (height drop) and flow rate with a weir board or flow meter.

Example: A 10-foot head and 10 gallons per minute flow rate can produce ~120 watts.

Turbine Selection:

Pelton Wheel: High-head, low-flow conditions.

Kaplan Turbine: Low-head, high-flow scenarios.

Crossflow Turbine: Versatile, ideal for medium-range sites.

Installation:

Build an intake system with debris filters.

Use PVC or HDPE pipes for penstock (the pipeline to the turbine).

Wire your turbine to an inverter and batteries for

usable AC power.

Tip: Install a diversion load to handle excess power and prevent turbine damage during low demand.

Off-Grid Living

Once you've generated energy, you'll need advanced systems for storing, managing, and conserving it.

BATTERY STORAGE SYSTEMS

Lithium-Ion: High upfront cost but long-lasting, lightweight, and efficient.

Lead-Acid: Affordable but heavier with shorter lifespans.

Saltwater Batteries: Eco-friendly but less energy-dense.

DIY Battery Bank Setup:

Connect batteries in series for higher voltage or parallel for more capacity.

Build a weatherproof battery enclosure. Add ventilation to prevent overheating.

Tip: Install a Battery Management System (BMS) to monitor voltage and prevent overcharging.

Backup Generators

Even the most advanced systems need a backup plan.

Dual-Fuel Generators: Flexible with gas and propane. Brands like Champion offer durable options.

Solar Generators: Silent and sustainable, great for emergency backup.

Calculating Energy Needs and Scaling Sustainably

Advanced homesteaders know the importance of matching energy supply with demand.

PERFORMING AN ENERGY AUDIT

List all appliances and devices with wattage and usage hours.

Calculate total daily energy needs (e.g., fridge: 150W x 24 hours = 3,600Wh/day).

Add a 25% buffer for inefficiencies and seasonal variances.

SCALING TIPS

Begin with essentials and expand as your energy needs or budget grow.

Use smart hybrid controllers from brands like Victron or Outback for seamless management of combined solar, wind, and hydro systems.

With advanced renewable energy skills, you're setting your homestead up for unmatched sustainability and resilience. Harness the sun, wind, and water, store and conserve energy wisely, and live with the confidence that your homestead is as self-sufficient as it is environmentally friendly.

CHAPTER SEVEN: FOOD STORAGE AND PRESERVATION

You've no doubt already mastered the foundations like canning and basic dehydration, and now it's time to explore advanced methods that will elevate your homestead pantry to new heights. In this chapter, you'll learn to build a root cellar, ferment like a master artisan, smoke and cure meats like an old-world butcher, and embrace cutting-edge long-term storage techniques such as freeze-drying and vacuum-sealing. Together, these methods will ensure your harvest lasts through lean seasons while providing creative ways to enhance the flavors of your bounty.

Building a Root Cellar: Nature's Refrigerator

A root cellar is more than just a storage area—it's a time-honored method for preserving fresh produce through natural refrigeration. Root cellars maintain optimal conditions for storing vegetables, fruits, and even canned goods for months without electricity.

CHOOSING THE RIGHT LOCATION

The first step is selecting a location with:

Consistent Cool Temperatures: Between 32–40°F (0–4°C).

High Humidity Levels: Around 85–95% to prevent dehydration.

Good Drainage: Ensure the area won't collect standing water.

Tip: A north-facing slope is ideal for maintaining a stable temperature, but if your land is flat, planting shrubs around your root cellar can provide shade and insulation.

Designing Two Root Cellars

The two types of root cellars to look into for your homestead are:

Underground Root Cellars:
Best for areas with stable, frost-free soil.
Requires excavation and reinforced walls.

Above-Ground Alternatives:
Ideal for high water tables or rocky terrain.
Build a mound cellar with straw bales and soil for insulation.

Below are two comprehensive plans: one for Underground Root Cellars and another for Above-Ground Alternatives, catering to different climates and soil conditions.

Underground Root Cellar Design

These root cellars are best for:
Areas with stable, frost-free soil.
Regions where the water table is low enough to prevent flooding.

Benefits:
Excellent thermal stability.
Long-lasting structure with minimal maintenance.
Can store a large quantity of produce.

Challenges:
Labor-intensive construction.
Not suitable for high water table areas.

Design Overview

An underground root cellar is built below ground level, leveraging the earth's natural insulation to maintain consistent temperatures and humidity levels year-round.

Materials Needed

Structural Components:
Concrete blocks or poured concrete for walls.
Pressure-treated wood or reinforced concrete for the ceiling.

Excavation Tools:
Backhoe or shovel for digging.

Ventilation System:
PVC pipes (4-inch diameter) for intake and exhaust.

Insulation:
Earth, straw bales, or foam insulation (optional for regions with extreme temperatures).

Door and Hatch:
Wooden or metal door, insulated to prevent heat exchange.

Construction Instructions

1. Site Selection and Preparation

Location: Choose a north-facing slope, if possible, to minimize sun exposure. Avoid areas prone to flooding.
Drainage: Ensure the site has proper drainage by installing gravel or a French drain system.

2. Excavation

Mark the dimensions of the cellar (e.g., 8x10 feet for a small homestead).

Dig down at least 6-8 feet to provide adequate insulation from ground temperature variations.

3. Foundation and Walls

Pour a 6-inch concrete slab at the base to prevent water ingress.

Build walls using concrete blocks or poured concrete.

Reinforce walls with rebar for stability.

Leave space for ventilation pipes at opposing ends.

Optional: Add a vapor barrier (plastic sheeting) beneath the slab.

4. Ceiling and Entrance

Install a reinforced concrete slab for the roof. Ensure it can support the weight of the overlying soil.

Create an entrance

Build a small staircase leading to a secure, insulated door.

For added insulation, add a secondary "airlock" door.

5. Ventilation System

Install one pipe near the floor for cool air intake.

Place another pipe at the ceiling level for warm air exhaust.

Extend both pipes outside, ensuring they are capped to prevent debris or pests from entering.

6. Cover with Earth

Backfill the top and sides with at least 2 feet of soil for insulation.

Plant grass or ground cover to reduce erosion.

7. Interior Setup

Add wooden shelves or bins for organized storage.

Maintain 70-90% humidity using damp sand or trays of water if necessary.

Above-Ground Mound Cellar

These types of root cellars are best for:

High water table regions.

Areas with rocky terrain where excavation is difficult.

Benefits

Easier and less labor-intensive to construct than an underground cellar.

Adaptable to challenging terrains.

Can be dismantled or relocated if needed.

Challenges

Less stable temperature compared to underground cellars.

Requires more attention to insulation and sealing.

Design Overview

Above-ground root cellars, or mound cellars, utilize insulation materials like straw bales and soil to mimic the conditions of an underground cellar.

Materials Needed

Structural Frame:

Wooden frame or cinder blocks.

Insulation:

Straw bales, soil, or a combination of both.

Roofing:

Plywood, metal sheeting, or corrugated panels.

Ventilation System:
PVC pipes (4-inch diameter) for intake and exhaust.

Entrance:
Wooden door with weatherstripping.

Construction Instructions

1. Site Selection and Preparation
Location: Choose a shaded area, preferably with natural windbreaks.
Foundation: Level the ground and spread a 4-6 inch layer of gravel for drainage.

2. Build the Frame
Construct a rectangular frame using treated wood or cinder blocks.
Dimensions: 8x10 feet is sufficient for a small cellar.
Height: At least 6 feet to provide ample storage space.
Ensure the frame is sturdy and level.

3. Add Insulation
Place straw bales or rigid foam insulation tightly against the walls.
Cover the exterior walls with a waterproof tarp or plastic sheeting to protect against moisture.

4. Roof Construction
Install a sloped roof using plywood or metal sheeting to shed rain and snow.
Add a layer of straw bales or soil on top for insulation.
Optional: Plant grass on the mound to enhance insulation and prevent erosion.

5. Ventilation System

Install one pipe at the base of the structure for cool air intake.

Place another pipe near the roof for warm air exhaust.

6. Entrance and Sealing

Build an insulated door at one end of the structure.

Weatherstrip the edges to minimize heat exchange.

Create a small vestibule or airlock to stabilize the internal temperature.

7. Interior Setup

Arrange bins or racks inside for produce storage.

Use thermometers and hygrometers to monitor temperature and humidity.

Check local zoning laws before digging to avoid underground utility lines.

Fermenting

Fermentation transforms your produce into nutrient-packed, flavorful staples.

Making Kombucha

Kombucha is a tangy, probiotic tea brewed with a SCOBY (Symbiotic Culture of Bacteria and Yeast). It's relatively easy to prepare once you get the steps down and there are many different flavor combinations to try.

What You'll Need

A gallon-sized glass jar.
A SCOBY and starter tea.
Sweet tea made from black or green tea and sugar.

Obtaining and Preparing SCOBY and Starter Tea for Kombucha

A SCOBY is the heart of kombucha brewing. This gelatinous, disc-shaped colony ferments sweet tea into the tangy, probiotic-rich drink. To begin brewing, you'll need a healthy SCOBY and starter tea, both of which play crucial roles in kickstarting the fermentation process.

Obtaining a SCOBY

Option A: Buy a SCOBY

SCOBYs are available online from reputable kombucha suppliers, at health food stores, or from local fermentation enthusiasts.

What to Look For:

A healthy SCOBY should be light beige to white, with no dark spots or mold.

It should come with 1-2 cups of starter tea for proper fermentation.

Option B: Grow Your Own SCOBY

If a SCOBY isn't readily available, you can grow one from scratch using store-bought, unflavored, raw kombucha.

Materials Needed:

1 bottle of raw, unflavored kombucha.
1 cup of sugar.
4 cups of water.
2 black or green tea bags.
A wide-mouth glass jar.

A cloth and rubber band.

Steps:

Brew sweet tea by dissolving sugar in hot water and steeping tea bags for 5-10 minutes. Allow the tea to cool to room temperature.

Pour the cooled sweet tea into the glass jar.

Add the raw kombucha to the jar as a "starter tea."

Cover the jar with the cloth secured by a rubber band to allow airflow while preventing contaminants.

Store in a warm, dark place (68°F-85°F or 20°C-29°C) for 1-4 weeks.

A SCOBY will begin forming on the surface as a thin film and thicken over time. Once it reaches about ¼ inch thick, it's ready to use.

Preparing the Starter Tea

Starter tea is fermented kombucha from a previous batch or the liquid that comes with a purchased SCOBY. It ensures the correct pH level to inhibit harmful bacteria during fermentation.

Why Starter Tea is Essential

Lowers the pH of the sweet tea, creating an acidic environment hostile to pathogens.

Introduces beneficial bacteria and yeast to jumpstart fermentation.

How to Prepare Starter Tea

If you have a purchased SCOBY, use the accompanying liquid as your starter tea.

If growing a SCOBY from scratch, the raw kombucha used in the process serves as your first starter tea.

Preparing the Sweet Tea for Brewing Kombucha

Sweet tea is the base for brewing kombucha.

Proper preparation ensures optimal conditions for the SCOBY and starter tea.

RECIPE FOR 1 GALLON OF KOMBUCHA

Ingredients

1 cup white cane sugar (organic if possible).
8 cups water
4-6 tea bags (black, green, or a combination). Avoid teas with oils or additives.
2 cups starter tea.

Instructions

Boil the Water: Bring 8 cups of water to a boil.

Steep the Tea: Add tea bags and steep for 5-10 minutes, depending on your desired strength.

Add Sugar: Stir in 1 cup of sugar while the tea is hot, ensuring it dissolves completely.

Cool the Tea: Allow the sweet tea to cool to room temperature. Hot tea can kill the SCOBY.

Combine with Starter Tea: Add 2 cups of starter tea to the cooled sweet tea.

Adding the SCOBY: Place the SCOBY gently into the sweet tea and starter tea mixture. It may float, sink, or sit at an angle—all are normal.

Cover the jar with a breathable cloth secured with a rubber band. Avoid using airtight lids, as the SCOBY needs airflow for fermentation.

Ferment for 7–10 days at room temperature. Taste regularly to find your desired level of tanginess.

Flavoring Ideas for Second Fermentation:

Ginger-Lemon: Add ginger slices and lemon juice for a zesty kick.

Berry Blast: Puree mixed berries and add to the jar.

Tips for Success:

Water Quality: Use filtered or distilled water to avoid chlorine or other chemicals that can harm the SCOBY.

Tea Choices: Stick to plain black or green tea for best results. Herbal teas or those containing oils (like Earl Grey) can disrupt the fermentation process.

Temperature: Keep your kombucha jar in a warm, stable environment (68°F-85°F or 20°C-29°C) but away from direct sunlight.

Benefits of DIY SCOBY and Starter Tea

Cost-Effective: Growing your own SCOBY eliminates the need for costly starter kits.

Sustainability: A single SCOBY can produce unlimited kombucha batches and grow new SCOBY layers to share or use in other recipes (like SCOBY candy or pet treats).

Customization: By creating your SCOBY, you gain greater control over the tea, sugar, and flavor choices in your kombucha.

With a healthy SCOBY and starter tea, you're ready to embark on your kombucha-brewing journey, crafting a delicious and healthful beverage perfectly tailored to your tastes.

Making Vinegar

Homemade vinegar serves as a preservative, cleaner, and cooking ingredient.

INGREDIENTS FOR VINEGAR

An alcohol base like wine or cider.

A mother of vinegar culture (available online or from previous batches).

INSTRUCTIONS

Mix alcohol with the mother culture in a wide-mouthed jar.

Cover with a cloth and store in a warm, dark place for 4–6 weeks.

Once tangy, strain and bottle the vinegar.

Tip: Save the "mother" for continuous batches, experimenting with different alcohols like mead for unique flavors.

Making Miso

Miso is a fermented soybean paste that brings depth to soups, marinades, and sauces.

INSTRUCTIONS

Soak and cook soybeans until soft.

Mash the soybeans and mix with salt and koji mold (available from fermentation supply stores).

Pack the mixture into an airtight container, leaving space for expansion.

Ferment for 6 months to 2 years.

Tip: Experiment with barley or chickpeas for different miso flavors.

Smoking and Curing Meats

Preserving meat with smoke and salt is a traditional art that delivers rich flavors and long shelf life.

HOT SMOKING VS. COLD SMOKING

Hot Smoking: Cooks and preserves at 225°F. Great for ribs, poultry, and fish.

Cold Smoking: Adds flavor without cooking

(under 90°F). Best for bacon and jerky.

STEPS FOR SMOKING MEAT

Brine the meat in saltwater for 24 hours to enhance flavor and moisture.

Use hardwoods like hickory or cherry for aromatic smoke.

Maintain a consistent temperature and monitor the meat's internal temp.

Dry Curing for Charcuterie:

Rub meat with a curing mix of salt, sugar, and spices.

Refrigerate for 1–2 weeks, flipping daily.

For advanced curing, hang the meat in a cool, humid room for aging.

Tip: Try curing pancetta or prosciutto for a gourmet touch.

Long-Term Food Storage

FREEZE-DRYING

Freeze-dried food retains its nutrients and lasts decades.

STEPS:

Use a home freeze dryer like Harvest Right to process fruits, vegetables, and even full meals.

Store freeze-dried items in Mylar bags with oxygen absorbers for extended shelf life.

VACUUM-SEALING

Vacuum-sealing extends the freshness of both dehydrated and fresh foods.

Tip: Seal bulk purchases or dried herbs to maintain potency. Pair with freezing for double protection.

DEHYDRATION

Elevate your dehydration game with solar-powered dehydrators or pre-treatments like blanching vegetables for better texture retention.

Fun Snacks to Try:
Spiced apple chips.
Jerky marinated in homemade vinegar.

With these advanced food storage and preservation techniques, your pantry will be a fortress of flavor, nutrition, and self-reliance. These methods not only preserve your harvest but also deepen your connection to the land and its cycles.

CHAPTER EIGHT: WILDCRAFTING AND FORAGING

Wildcrafting and foraging are where science meets art, connecting homesteaders to the natural abundance of their surroundings. These skills go beyond gathering food and delve into the realms of sustainability, ecological stewardship, and advanced self-reliance. By incorporating wild foods and medicinal plants into your homestead, you not only diversify your food sources but also enrich your life with seasonal rhythms and a deeper connection to nature.

Identifying Wild Edibles

Foraging is hyper-local. Even if you've mastered the basics, understanding the nuances of wild edibles in your area is key to advancing.

TEMPERATE ZONES

Wild Roots: Burdock (harvest in early spring or late fall for tender roots).

Tip: Burdock root can be pickled or added to stews for a sweet, earthy flavor.

Greens and Shoots: Stinging nettle (best in early spring).

Tip: Wear gloves and blanch nettles to neutralize their sting. They're packed with iron and perfect for soups or teas.

TROPICAL ZONES

Fruits: Breadfruit and wild tamarind.

Tip: Breadfruit, when roasted and mashed, can be a starchy substitute for potatoes or flour.

Leaves: Moringa is a nutritional powerhouse

growing wild in many tropical areas.

DESERT AREAS

Edible Succulents: Harvest fleshy pads of prickly pear cacti. Remove spines carefully with a knife and use them in stir-fries or as a taco filling.

COLD/ALPINE CLIMATES

Fungi: Understand the habitat of high-elevation mushrooms like the matsutake, which fetch premium prices in gourmet markets.

Tip: Matsutakes grow near specific tree species such as pines or firs. Learn to identify their natural companions.

Multi-Sensory Identification Skills

Sight: Look for defining features like leaf arrangement, flower shapes, or bark texture.

Smell: Some wild foods, like mint or garlic mustard, are unmistakable when crushed.

Touch: Textures can differentiate plants. Lamb's ear has a velvety feel, unlike most toxic plants.

Taste Testing Safely: Once confident in identification, taste a tiny sample, as many wild edibles have intense flavors or textures.

Learning Plant Relationships

Research symbiotic relationships between plants, fungi, and soil types to predict where certain edibles will thrive.

Truffles: Found under oak or hazelnut trees due to their symbiotic relationship.

Berries: Look for wild strawberries in sunny, well-drained soils near forest edges.

Wildcrafting Medicinals

Understanding the medicinal power of plants can turn your homestead into a healing sanctuary.

Goldenrod: An advanced forager's ally for urinary health and allergies.
How to Use: Dry the flowers and leaves to make teas or tinctures. Goldenrod salves can also reduce inflammation.
Boneset: Excellent for treating fevers.
Tip: Brew into a tea, but avoid overuse as it can act as a laxative.

Harvesting

Timing is Key: Medicinal potency often peaks at certain times of day or growth stages. Harvest flowers in full bloom and roots in the dormant season.

Making Potent Preparations

Invest in airtight jars, amber bottles, and a dehydrator for preserving medicinal herbs. Label everything with harvest dates and usage notes for best practices.

INFUSED OILS

Recipe: Gather fresh St. John's wort. Pack a jar and cover with olive oil. Leave in the sun for several weeks to infuse. Use the red-tinted oil for burns and muscle pain.

DUAL-EXTRACT TINCTURES FOR MUSHROOMS

Use alcohol and water in separate steps to extract both alcohol-soluble and water-soluble compounds. Reishi and chaga mushrooms are ideal for this process.

Sustainable Foraging Practices

ETHICS OF ADVANCED WILDCRAFTING

Plant wild seeds from foraged herbs around your property to encourage natural regeneration.

Leave Seed Plants: Focus on harvesting leaves, stems, or flowers, leaving roots and seeds to regenerate the plant.

Harvest with Purpose: Gather only what you'll use. This prevents waste and ensures availability for others.

Foraging Tools

Foraging successfully and sustainably requires a mix of tools to harvest plants, mushrooms, and other wild foods while respecting the environment. Here's a detailed list of foraging tools and their uses, divided into categories for better organization.

GATHERING AND HARVESTING TOOLS

Harvesting Bags: Collecting plants, fruits, mushrooms, or nuts while allowing air circulation to prevent spoilage.

Types:

Mesh Bags: Ideal for seeds, berries, and small fruits, allowing seeds to scatter naturally.

Canvas Totes: Durable and reusable for heavier items like roots or larger mushrooms.

Forager's Basket: Sturdy and breathable, perfect

for delicate mushrooms or berries.

Soil Knives (Hori Hori Knives): Digging up roots, dividing plants, and cutting through tough soil or stems.
Features: Serrated edge for sawing and a pointed tip for digging.

Pruners or Shears: Trim herbs, branches, or tough stems without damaging the plant.
Tip: Use bypass pruners for clean cuts that promote plant regrowth.

Folding Saw: Cut thicker branches or saplings for foraged wood or materials.
Benefits: Compact design makes it easy to carry.

Berry Picker Tool: Collect small berries like blueberries or lingonberries quickly and efficiently.
Design: Often features a comb-like scoop to strip berries from branches.

DIGGING AND CUTTING TOOLS

Trowel: Remove soil when harvesting underground tubers or bulbs like wild garlic and dandelion roots.

Handheld Hoe or Mattock: Break up compact soil to access deeper roots or tubers.

Mushroom Knife: Specially designed for harvesting mushrooms with a curved blade for clean cuts.
Additional Feature: Many include a brush on the handle for cleaning dirt off mushrooms.

Multi-Tool: A versatile tool with knife blades,

saws, screwdrivers, and other functions for various foraging tasks.

TRANSPORT AND STORAGE TOOLS

Paper Bags or Waxed Paper: Store mushrooms or greens without causing moisture buildup, which can lead to spoilage.

Plastic Buckets or Collapsible Containers: transport larger or bulkier items like nuts, fruits, or large mushrooms.

Belt Pouches or Waist Bags: Keep small tools, gloves, or identification guides handy while leaving hands free.

SAFETY AND IDENTIFICATION TOOLS

Foraging Gloves: Protect hands from thorny plants, nettles, or rough bark.

Types:

Leather: Durable and protective for heavy-duty tasks.

Nitrile-Coated: Lightweight and better for delicate handling of plants.

Magnifying Glass or Loupe: Examine plant leaves, fungi gills, or small details to aid in identification.

Field Identification Guides: Ensure accurate identification of plants, fungi, or berries.

Digital Mapping Apps: Mark high-yield spots, track seasonal changes, and prevent getting lost.

Examples: Gaia GPS, iNaturalist, or Avenza Maps.

First Aid Kit:

PRESERVATION AND PROCESSING TOOLS

Portable Dehydrator (Solar or Electric): Dry mushrooms, herbs, or fruits immediately after foraging to preserve freshness.

Scissors: Clip herbs or small plants precisely without damaging surrounding growth.

Paring Knife: Clean and trim fruits, roots, or vegetables on-site if needed.

Hand-Powered Grinder or Mortar and Pestle: Grind nuts, seeds, or herbs during or after foraging.

SPECIALIZED TOOLS

Nutcracker or Pliers: Open tough shells of foraged nuts like walnuts or hazelnuts.

Fishing Net: Harvest aquatic plants like cattails or watercress.

Clippers with a Hooked Blade: Reach and cut hard-to-access fruits or branches.

Rake or Long-Handled Grabber: Collect fallen nuts or fruits under trees without bending down repeatedly.

CLOTHING AND ACCESSORIES

Wide-Brimmed Hat: Protect against sun exposure during long foraging trips.

Waterproof Boots: Navigate wet or muddy areas safely.

Apron with Pockets: Carry small tools or

harvested items conveniently.

SUSTAINABILITY TOOLS

Seed Spreader: Scatter seeds of native plants to promote regrowth and preserve ecosystems.

Compostable Trash Bags: Collect any trash found during foraging to leave the environment cleaner than you found it.

Marking Flags or Tape: Mark areas for sustainable harvesting, ensuring that plants have time to regrow between visits.

TIPS FOR FORAGING WITH THESE TOOLS

Lightweight Carrying: Invest in a sturdy backpack with multiple compartments to distribute weight evenly.
Multipurpose Tools: Opt for tools that serve more than one function, like a mushroom knife with a cleaning brush.
Seasonal Adjustments: Tailor your tool kit to what you're foraging for—different seasons and regions may require variations.
Clean After Use: Always clean tools after use to prevent contamination or the spread of invasive species.

With these tools, you'll be prepared for any foraging adventure, whether you're gathering roots, harvesting wild greens, or identifying edible mushrooms.

Foraging for Livestock

WILD EDIBLES FOR ANIMAL FEED

Goats and Sheep:

Collect mulberry leaves, a nutrient-dense feed that encourages weight gain and milk production.

Poultry:

Add forage greens like clover and purslane for omega-3-enriched eggs.

LIVESTOCK-SPECIFIC WILD FOODS

Some plants serve dual purposes, offering nutrition for both humans and animals.

Pig Forage: Nuts like acorns and chestnuts can be dried and stored for winter feed.

ADVANCED GRAZING STRATEGIES

Integrated Foraging Zones: Allow livestock to graze in areas rich in wild edibles, promoting natural behaviors while reducing feed costs. Rotate animals frequently to prevent overgrazing.

Section 5: Turning Wild Foods into Culinary Delights

1. Wildcrafted Staples

Pine Needle Tea: High in vitamin C, a simple brew from pine needles boosts immunity during cold months.

Nettle Chips: Dry or bake nettle leaves with olive oil and salt for a nutritious snack.

2. Preserving the Harvest

Freezing Berries: Lay wild blueberries or blackberries on a tray to freeze individually before storing in airtight containers.

Drying Mushrooms: Slice and dry chanterelles or boletus mushrooms for soups and stews year-round.

3. Recipes That Shine

Acorn Pancakes:

Soak and leach tannins from acorns, then grind

into flour. Mix with eggs, milk, and honey for fluffy, nutty pancakes.

Wildflower Jellies:

Harvest edible flowers like violets or dandelions. Use pectin and sugar to create jellies that capture the essence of spring.

Additional Resources for Mastery

After years of research I crafted food forest design guides based on every single planting zone in the United States, Canada and Europe. These guides include 10 different plans per zone and are invaluable resources for planting what will flourish in each zone, ensuring abundant food forests for generations. Check them out on Amazon if you're looking for a deeper dive into crafting, planting and layout design of your very own food forest!

Mastering wildcrafting and foraging is more than just collecting nature's bounty—it's about understanding your ecosystem and becoming an active participant in its cycles. With these advanced techniques, you'll elevate your homestead's sustainability and create a richer connection to the land, all while reaping its delicious and medicinal rewards. Nature always has more to offer when you learn to look deeper.

CHAPTER NINE: WATER MANAGEMENT SYSTEMS

Water sustains life, and on your homestead, it's the critical resource that underpins everything from growing food to keeping livestock hydrated. In this chapter, we'll take a deep dive into advanced water management systems that go beyond the basics. Whether it's recycling water through greywater systems, digging a well to access groundwater, creating multi-functional ponds, or purifying water for potable use, mastering these techniques will make your homestead a beacon of sustainability and resource efficiency.

Creating a Greywater System

Greywater systems are an advanced method of reusing water from sinks, showers, and laundry to irrigate plants, reducing water waste and cutting utility costs. This system is especially useful in regions prone to drought or for homesteaders seeking maximum resource efficiency.

WHY GREYWATER SYSTEMS:

Water Conservation: Reduces reliance on freshwater for irrigation.

Cost-Effectiveness: Lowers water bills and maximizes resource usage.

Eco-Friendliness: Supports sustainability by recycling non-potable water.

KEY COMPONENTS OF A GREYWATER SYSTEM

Source Points: Collect water from showers, bathtubs, and washing machines using only biodegradable soaps and detergents. Avoid collecting water from kitchen sinks due to grease and food particles (blackwater).

Filtration: Gravel, sand, or mesh filters help remove particles before water is distributed.

Diversion System: Install diverter valves to direct water to the garden or back to the sewer as needed.

Distribution: Use perforated pipes buried under mulch or gravel to distribute water directly to plant roots.

HOW TO BUILD A DIY GREYWATER SYSTEM

Assess Legal Compliance: Research local codes; some areas may require permits.

Install Diverters: Use three-way valves to switch water flow between sewer and greywater system.

Create Mulch Basins: Dig shallow trenches around fruit trees or shrubs and fill with mulch to absorb water and nutrients.

Add Basic Filters: Construct a gravel or charcoal drum filter for laundry or shower water.

TIPS FOR OPTIMIZING GREYWATER SYSTEMS

Rotate watering zones to prevent salt buildup in soil.

Avoid using greywater on root vegetables or leafy greens to minimize health risks.

Regularly clean filters and check for clogs in the distribution pipes.

Digging and Maintaining Wells

Accessing groundwater through a well offers a steady, reliable water source that doesn't depend on rainfall. While this is a larger investment of time and resources, it's one that will serve your homestead for decades.

PLANNING YOUR WELL

Research Water Tables: Contact your local geological survey or use online resources like USGS (United States Geological Survey) to locate aquifers.

Soil Testing: Send soil samples to a lab to assess permeability and composition. Sandy soils are ideal; rocky or clay soils may require more intensive equipment.

Permit Acquisition: Research local regulations and secure necessary permits.

DIGGING THE WELL

Shallow Hand-Dug Wells: Use a shovel and a pulley system for water tables less than 25 feet deep. Line the walls with bricks or pre-cast concrete rings to prevent collapse.

Drilled Wells: Rent or purchase an auger drill to access deeper water. These are more suitable for aquifers 50–200 feet below the surface.

Install the Pump: Choose a pump based on your setup:

Manual Pumps: Simple and off-grid friendly.

Electric Pumps: Efficient for high-volume needs; pair with solar panels for sustainability.

MAINTAINING YOUR WELL

Clean the well annually with a submersible pump to remove sediment.

Test water for contaminants like bacteria, nitrates, and heavy metals. Home testing kits or professional services can be used.

Protect the wellhead by sealing any cracks with grout to prevent surface water contamination.

Tip: For off-grid homesteads, consider a windmill-driven water pump, a classic yet effective solution for continuous water access.

Building Ponds

Ponds are versatile additions to any homestead, functioning as irrigation reservoirs, aquaculture habitats, and water features that enhance biodiversity.

DESIGNING YOUR POND

Purpose: Decide whether your pond will prioritize irrigation, aquaculture, or both.

Placement: Select a low-lying area that naturally collects runoff. Avoid areas close to septic systems to prevent contamination.

Size and Depth:

For irrigation: Rectangular ponds with sloping sides are easiest to manage.

For aquaculture: Incorporate deep sections (6–8 feet) for overwintering fish and shallow areas for spawning.

Step 2: Constructing the Pond

Use an excavator or backhoe to dig. For larger projects, consider hiring a contractor.

Line the pond with clay, bentonite, or a durable pond liner to prevent leaks.

Install an overflow pipe to manage excess water during heavy rains.

Step 3: Adding Aquaculture

Fish Stocking: Tilapia, catfish, or trout are excellent choices for homestead ponds.

Aquatic Plants: Include species like duckweed or cattails to oxygenate the water and support fish health.

Ecosystem Management: Regularly check oxygen levels and prevent algae overgrowth by balancing plant life.

Pro Tip: Use a gravity-fed irrigation system to channel pond water to your garden. Solar-powered pumps can assist with uphill distribution.

Filtering and Purifying Water Naturally

Ensuring access to potable water is essential, even for the most advanced homestead. Natural filtration systems allow you to clean water without chemicals or heavy reliance on commercial systems.

Techniques for Natural Water Purification

Bio-Sand Filters: Create a barrel-based filter with layers of gravel, sand, and activated charcoal. The top layer of sand houses bacteria that digest contaminants.

Constructed Wetlands: Build a wetland area with aquatic plants to filter greywater or pond runoff. These systems work as living filters, removing toxins and excess nutrients.

Solar Disinfection (SODIS): Fill clear plastic bottles with water and leave them in direct sunlight for 6–8 hours. UV rays kill most bacteria and viruses.

Building a DIY Bio-Sand Filter

Start with a large plastic drum.

Add a bottom layer of coarse gravel (4–6 inches), followed by fine sand (12–16 inches), and top it with activated charcoal (2 inches).

Attach a spout at the bottom for collecting filtered water.

Tips for Water Safety

Pre-filter water to remove sediment before advanced filtration.

Test your system periodically using portable water quality kits, available from suppliers like Sawyer or Berkey.

The USGS Website

If you happen to live in the United States, a resource that is invaluable, especially for planning and designing water systems on your homestead is the United States Geological Survey (USGS) website.

It is a treasure chest of information for planning your homestead, especially when it comes to understanding your land and water resources. I want to break the site down for you so you can familiarize yourself—if you're not already.

Imagine you're building a castle, and you need to know where the best rivers, wells, and soil are to make it strong. The USGS helps you do exactly that for your homestead! It has maps, data, and tools to help you understand the land under your feet and the water around you.

You can use it to find:

Where water flows on your property.

What kind of soil you have.

If your area is prone to flooding or drought.

Places to dig a well or collect rainwater.

WATER MAPPING

Knowing where water comes from and where it goes on your land is key for farming, gardening, and sustainability. Here's how you can use the USGS website to map water on your homestead:

Step 1: Go to www.usgs.gov.

Once there, look for the "Science" tab at the top or use the search bar to find specific tools like "Water Resources."

Step 2: Navigate to the National Water Dashboard

Type "National Water Dashboard" into the search bar, or go directly to the Water Resources section. Alternatively you can go to: **https://dashboard.waterdata.usgs.gov/** and pull up the map.

This tool shows you real-time data on rivers, lakes, and groundwater near your location.

Step 3: Search for Your Location

On the National Water Dashboard, find the map and type your city, state, or zip code into the search bar.

The map will zoom in on your area and show:

Rivers and streams.

Groundwater levels.

Rainfall data.

Step 4: Look at Streamflow Data

On the map, click on the blue dots representing water measurement stations.

These stations provide information about:

How much water is flowing in nearby rivers.

If the stream levels are high or low, which helps in planning for irrigation or preventing floods.

Step 5: Check Groundwater Levels

Back on the main page, search for "Groundwater Watch."

Here, you can find wells near your area and learn how deep the water is underground.

If you want to drill a well, this tool helps you choose the best spot.

Step 6: Study Flood and Drought Maps

On the site, look for "Floods and Droughts" under Water Resources.

Use these maps to see if your homestead is at risk of flooding or if the area has water shortages.

This helps you decide if you need to build drainage ditches or install large rainwater tanks.

Step 7: Soil and Land Data

Go to the Earth Resources Observation and Science Center (EROS) section for detailed soil maps.

Soil maps show you what kind of ground you have:

Sandy soil (good for drainage).
Clay soil (holds water well).
Loamy soil (perfect for gardening)

HOW DOES THIS HELP MY HOMESTEAD?

Smart Water Use: You'll know where to collect rainwater, build ponds, or set up irrigation systems.
Well Placement: Learn the best spots to drill for groundwater.
Flood Prevention: Build swales, berms, or drains to protect your home and crops from water damage.
Drought Planning: Identify areas for water storage and drought-resistant plants.
Efficient Gardening: Choose crops and plant locations based on soil type and water availability.

Mastering water management transforms your homestead into a sustainable and resilient system. These advanced methods ensure that every drop of water contributes to your self-sufficiency, from reusing greywater to supporting aquaculture and purifying drinking water. With these tools, you're not just surviving—you're thriving in harmony with nature.

CHAPTER TEN: INFRASTRUCTURE PROJECTS

As your homestead transforms from a humble plot into a thriving, self-sufficient sanctuary, the next logical step is to focus on building a robust infrastructure that can support your growing needs. A homestead's success lies not only in its daily operations but also in the durability, design, and flow of its physical elements—fences, barns, outbuildings, pathways, and functional zones. Think of these structures as the bones of your homestead: the sturdier and better planned they are, the more resilient and sustainable your homestead will be.

In this chapter, we'll dive into the specifics of advanced homestead infrastructure projects, from building fences that can handle all kinds of livestock and weather conditions, to creating pathways that guide you seamlessly through your property. We'll explore natural building techniques to help you construct durable and energy-efficient buildings using sustainable materials, and even show you how to tackle some of your own plumbing and electrical systems, giving you more independence and control.

Building Fences, Barns, and Outbuildings

FENCING

Fences are often the first project a homesteader tackles, but as you move into the realm of advanced homesteading, fencing becomes about more than just marking boundaries. It's about creating a system that works for your land, animals, and aesthetic. Advanced fencing involves choosing materials based on both function and durability and building with an eye for longevity.

Choosing the Right Type of Fence

Wire Fencing: Think beyond the basic chicken wire. Use welded wire or hog panels for stronger, more secure enclosures. These materials are ideal for goats, sheep, and even pigs, and they're easy to install. For added predator protection, install a layer of electric wire along the top or bottom to deter climbing or digging animals like raccoons and coyotes.

Wooden Fencing: If you're working with larger animals like cattle or horses, go for heavy-duty wood fences. Pressure-treated posts are critical for resisting rot, especially in areas with high moisture levels. Cedar and redwood are naturally resistant to decay and insects, making them ideal choices. For extra stability, add cross-bracing to the corners and gates to prevent sagging.

Living Fences: For a sustainable and aesthetically pleasing alternative, consider planting a living fence. Shrubs like hawthorn or blackberries, or even trees such as elderberry or osage orange, can form dense, thorny hedges that create both a boundary and wildlife habitat. Over time, these grow into natural barriers that are hard for animals to breach.

Building Tips:

Post Installation: To ensure your fence stands the test of time, dig post holes deep enough—at least 2 feet for sturdy wooden posts, and deeper in loose or sandy soil. The posts should be anchored with concrete or gravel at the bottom to prevent shifting. Consider using natural posts, like cedar, for long-lasting, chemical-free options.

Gate Construction: A strong gate is vital. Consider using solid wooden gates with heavy-duty hinges for strength. If possible, install your gate to swing both ways so you never get stuck while moving livestock or equipment.

Predator-Proofing: Small predators like foxes and raccoons can dig under fences. A smart solution is to bury wire mesh 6–12 inches underground or extend a trench below the fence to create a solid barrier.

Barns and Outbuildings

A barn isn't just a place to store hay or house animals—it's the central hub of activity on the homestead. Building and maintaining your barn (or multiple outbuildings) properly can transform your homestead's efficiency.

PLANNING YOUR BARN

Think about what your barn needs to do: Will it house animals and store tools? Is it going to double as a workshop or provide space for milking? If you're expanding your livestock operation, consider adding isolation pens, quarantine spaces, or even a milking parlor.

Designing for Efficiency:

Zones: Organize the barn into separate zones. You may have one section for feed storage, another for equipment, and a space for animals to rest. This reduces clutter and makes your day-to-day chores much easier.

Ventilation and Light: Natural light and airflow are key, especially when you're housing animals. Skylights or high windows allow light to flood the barn, while vents along the roofline help humidity escape. Using solar-powered lights is a cost-effective way to brighten the space without wiring in electricity.

Natural Building Materials for Barns:

Cob and Earth: Cob construction (a mixture of

clay, sand, and straw) has been used for thousands of years. It's energy-efficient and has excellent thermal mass—keeping your barn cool in the summer and warm in the winter.

To build with cob, start by mixing clay, sand, and straw in the right proportions. A typical ratio is 1 part clay to 3 parts sand, with straw making up about 1/3 of the mix. Once you've created your cob, it's simply a matter of layering and stamping it into place. After allowing it to dry and cure, your structure will be incredibly durable and energy-efficient.

Straw Bale: Straw bale construction is another effective way to build durable, well-insulated barns or outbuildings. Straw is often free or very inexpensive, and it has a high R-value, making it an excellent insulator.

Stack the bales tightly together, and plaster over them with a lime and sand mixture to protect against the elements. Straw bale buildings are also natural fire retardants, thanks to the density of the bales.

Earthbags: Earthbag construction is a unique and sturdy way to build a homestead structure. Earthbags (woven polypropylene bags filled with soil) are stacked like bricks to form walls. This technique works particularly well in areas with seismic activity or where soil is abundant.

Earthbag construction is highly effective in creating structures that are naturally resistant to earthquakes and floods.

Creating Pathways and Functional Zones

A homestead is much more than a collection of separate projects—it's a living, breathing system. Pathways and functional zones are the arteries and veins that connect everything, making it easy to navigate, access, and maintain your space.

DESIGNING PATHWAYS

Pathways should serve a functional purpose while also complementing your homestead's aesthetic. Whether you need to make a trek from the house to the barn, or from your vegetable garden to your compost pile, a well-placed pathway saves time, effort, and wear on your shoes.

Types of Pathways:

Gravel Paths: Gravel is a durable, inexpensive material that provides excellent drainage. When installing a gravel path, ensure a thick enough layer (at least 3 inches) and compact it to prevent shifting.

Wooden Boardwalks: If you live in a damp environment, wooden boardwalks are perfect for keeping your feet dry while walking through wet areas. Raised above the ground, they allow air circulation and prevent the buildup of mold.

Mulch or Wood Chips: For an eco-friendly, low-maintenance pathway, mulch and wood chips are ideal. These materials not only create a soft surface for walking but also decompose slowly, enriching your soil with organic matter.

Design Tips:

Function First: Plan your pathways around your most frequently used areas—your barn, garden, and tool sheds. Think about the paths that will save you time during your daily routines.

Aesthetic Touches: While functionality is key, pathways can also beautify your space. Add flower beds, native plants, or little benches along the way for a welcoming touch.

Functional Zones

To maximize your homestead's efficiency, organize your land into functional zones. This zoning makes it easier to manage your space and reduces the amount of time spent walking from one area to another.

Zone 1 (Immediate Zone): This is your high-maintenance, frequently visited area. Place your herb garden, vegetable garden, and composting bins here, as these are things you'll want to be close to.

Zone 2 (Animal Zone): The animal zone should be close to the house for ease of access, but separate enough to allow for safe pastures and free-range zones. Ensure this space is predator-proof and allows for rotation of grazing animals.

Zone 3 (Productive Zone): This is your larger orchard, greenhouse, or food forest. Though it's a bit farther from the house, it's still accessible for regular maintenance and harvesting.

Zone 4 (Wilderness Zone): This area is for off-grid activities, firewood storage, or simply preserving natural habitat.

Natural Building Techniques

Embrace the art of sustainable building with natural materials—cob, straw bale, and earthbag construction. These methods use readily available materials to create homes and outbuildings that are both durable and eco-friendly.

COB CONSTRUCTION

Cob is an ancient technique that blends clay, sand, and straw to create strong, energy-efficient buildings.

The natural ingredients create walls that are not only highly insulating but also incredibly durable.

Building with Cob:

Mix clay, sand, and straw in the right proportions (about 1 part clay to 3 parts sand).

Layer the cob mixture in small sections, stamping each layer down tightly.

Use your feet to compact the mixture or a tamper for better results. This method produces walls that have excellent thermal mass.

STRAW BALE CONSTRUCTION

Straw bale is another excellent building method, especially for insulating buildings in both hot and cold climates. Straw is inexpensive and readily available, making it an affordable option.

Building with Straw Bales:

Stack the straw bales tightly, ensuring each one is compressed and level.

Plaster over the bales with clay or lime to protect them from the elements and pests.

EARTHBAG CONSTRUCTION

Earthbag construction involves filling bags with soil and stacking them to form walls. These structures are durable, weather-resistant, and offer excellent insulation properties.

Building with Earthbags:

Fill polypropylene bags with soil and stack them to create walls. Compact each layer thoroughly.

Earthbag construction is great for building structures in earthquake-prone regions, as the walls are highly resistant to shifting.

DIY Plumbing and Electrical Projects

As your homestead grows, so will your need for plumbing and electricity. By mastering DIY plumbing and electrical systems, you can achieve greater self-sufficiency and reduce your reliance on external utilities.

DIY PLUMBING

Start small with simple systems, such as rainwater harvesting or greywater reuse, and build up to more complex setups.

Rainwater Harvesting: Install gutters and downspouts to collect rainwater. Use a first-flush diverter to keep debris out, and store the water in large barrels or a cistern.

Greywater Systems: Install a simple system to redirect water from sinks, showers, and laundry to irrigate your garden.

DIY ELECTRICAL

If you're looking to go off-grid, solar, wind, and hydroelectric power are the most common DIY energy sources.

Solar Power: Install photovoltaic panels and connect them to an inverter to convert the DC current into AC power.

Wind Power: Small-scale wind turbines can provide consistent power in areas with good wind resources.

Hydroelectric Power: A micro-hydropower system can work wonders if you have access to a nearby stream or river.

Below are four functional design and construction plans for each zone of your homestead, incorporating cob, straw bale, or earthbag construction.

Zone 1: Cob Garden Shed with Integrated Compost System

Purpose: A multipurpose garden shed for tool storage, seed organization, and compost integration.

MATERIALS:

Clay soil: 1 cubic yard (ensure it's subsoil with no organic matter).
Sand: 2 cubic yards (sharp sand for structural integrity).
Straw: 4 bales (acts as a binder).
Water: 150–200 gallons (depending on soil moisture).
Reclaimed wood: 6 pieces, 2"x4"x8' for framing the door.
Glass or clear plastic: For windows or a small skylight
Gravel: 1 cubic yard for foundation drainage.
Wire mesh: 10 square feet (optional for extra tensile strength in high-stress areas).

TOOLS:

Shovel
Wheelbarrow
Cob stomping tarp
Level
Trowels
Saw (for wood cutting)
Tape measure
Hammer and nails

INSTRUCTIONS:

Foundation:
Excavate a 12" trench for the 8'x8' shed footprint.

Fill the trench with 6" of gravel for drainage.

Cob Mixing:
On a tarp, mix 1 part clay soil, 2 parts sand, and water until it's a dough-like consistency.
Add straw and mix further to create a strong, fibrous mixture.

Wall Building:
Layer cob in 6" increments, allowing each layer to partially dry before adding the next.

Leave spaces for windows and the doorframe.

Roof:
Use a simple gable roof with reclaimed wood beams and corrugated metal or reclaimed roofing panels.

Integrated Compost System:

Build a recessed compartment into one side of the shed for compost bins.
Include ventilation holes and a waterproof lining.

Zone 2: Straw Bale Chicken Coop

Purpose: A predator-proof chicken coop with natural insulation to keep birds comfortable year-round.

MATERIALS:

Straw bales: 20 large bales.
Wood planks: 10 pieces, 2"x4"x8' for framing.
Chicken wire: 50 feet.
Earth plaster: Mix of 1 part clay, 2 parts sand, and water (5 gallons per plaster layer).
Corrugated roofing panels: 4 panels for a 6'x8' coop.
Gravel: 1 cubic yard for the foundation.
Metal mesh: For a predator-proof floor.

TOOLS:

Staple gun
Wire cutters
Shovel
Level
Trowel
Drill with screws
Mallet
Wheelbarrow

INSTRUCTIONS:

Foundation:
Clear a 6'x8' area and dig a shallow trench.

Fill with 4" of gravel to ensure good drainage.

Straw Bale Walls:
Stack bales in a running bond pattern, alternating the direction of bales on each layer.

Secure bales to the wood frame using wooden stakes or metal rods.

Roof:
Frame a simple sloped roof with the 2"x4" planks.

Attach corrugated panels for weatherproofing.

Chicken Access and Security:

Cover all windows with chicken wire.

Place a predator-proof metal mesh under the coop floor.

Earth Plaster:
Apply a layer of earth plaster to the exterior for weather resistance.

Zone 3: Earthbag Greenhouse

Purpose: A sturdy, thermally insulated greenhouse for extended growing seasons.

MATERIALS:

Earthbags: 300 bags (polypropylene or burlap).
Clay soil: 2 cubic yards.
Sand: 1 cubic yard.
Gravel: 2 cubic yards.
Barbed wire: 200 feet (acts as a "Velcro" layer between earthbag rows).
Reclaimed wood or metal frame: For door and window frames.
Plastic sheeting or polycarbonate panels: For the greenhouse covering.
Rebar: 20 pieces, 4' each for structural stability.

TOOLS:

Tamper

Bucket
Shovel
Level
Wire cutters
Rubber mallet
Tarp for mixing

INSTRUCTIONS:

Foundation:
Excavate an 8’x12’ area to a depth of 12”.

Fill with gravel to prevent water seepage.

Earthbag Walls:
Fill each bag with a mix of clay, sand, and a small amount of water.

Lay the bags in rows, tamping each layer flat.

Place barbed wire between rows for stability.

Greenhouse Covering:
Build a curved metal or wood frame over the structure.

Attach clear plastic sheeting or polycarbonate panels.

Ventilation and Doors:
Install vents and a sturdy doorframe.

Zone 4: Cob Firewood Shelter

Purpose: A rustic cob structure to keep firewood dry and organized in the wilderness zone.

MATERIALS:

Clay soil: 1 cubic yard.
Sand: 2 cubic yards.
Straw: 3 bales.
Reclaimed wood: For a roof frame and shelving.
Gravel: 1 cubic yard for drainage.
Metal roofing sheets: 4 sheets for a sloped roof.

TOOLS:

Shovel
Wheelbarrow
Trowels
Saw
Level
Tape measure

INSTRUCTIONS:

Foundation:
Clear a 5'x10' area and dig a trench 12" deep.

Fill with gravel for drainage.

Cob Walls:
Build 2'-thick cob walls, allowing for gaps to improve airflow.

Roof:
Frame a simple sloped roof with reclaimed wood.

Attach metal sheets for waterproofing.

Shelving:
Build interior shelving using reclaimed wood to stack firewood neatly.

These projects blend functionality with sustainability, showcasing how cob, straw bale, and earthbag methods can enhance each homestead zone while utilizing eco-friendly techniques.

By mastering these advanced infrastructure projects, you'll unlock the full potential of your homestead, ensuring it's not just self-sufficient but also a sustainable, resilient place to live for generations to come.

CHAPTER ELEVEN: CREATING A HOMESTEAD RETREAT

A thriving homestead is not just a testament to hard work and resilience—it's also a place to rest, reflect, and recharge. Beyond the cycles of planting, harvesting, and tending, creating a retreat on your homestead is essential for your physical and mental well-being.

Nature-Based Wellness Practices

FOREST BATHING

Deep Connection with Your Land

Forest bathing (*shinrin-yoku*) isn't just for large wooded areas—it can be adapted to your homestead, even if your property is modest.

Identify a quiet area with trees or shrubs, or cultivate a "forest patch" using fast-growing species like poplar, alder, or willow.

Plant understory shrubs such as hazelnut or elderberry to mimic natural forest layers.

Engaging the Senses:

Sight: Observe sunlight filtering through leaves. Focus on the textures of bark or the movement of branches in the breeze.

Sound: Listen for the subtleties—birdsong, rustling leaves, or distant flowing water.

Smell: Crush a leaf or pinch an herb between your fingers to release its aroma.

Touch: Feel the coolness of moss or the rough texture of tree bark.

Tip: Document the flora and fauna you

encounter during your sessions. This can double as a mindfulness exercise and an opportunity to monitor your property's biodiversity.

Outdoor Yoga and Meditation

Yoga and meditation in nature combine the physical benefits of movement with the grounding effects of your land.

CREATING A DEDICATED SPACE

Build a simple platform using reclaimed wood. Ensure it's level and placed in an area with dappled shade.

Surround it with aromatic plants like lavender, lemon thyme, or mint to stimulate the senses during practice.

Seasonal Practices:

Spring: Energizing sun salutations to greet longer days.

Summer: Cooling breaths (sheetali pranayama) in the early morning or evening.

Fall: Gentle twists to release tension and prepare for winter.

Winter: Meditate near a fire pit or in a greenhouse to stay connected to nature despite the cold.

Tip: Incorporate elements from your land into your practice. For example, hold a smooth stone during meditation or practice yoga poses inspired by local wildlife (e.g., tree pose under an actual tree).

Designing Sacred Spaces

MEDITATION GARDENS

Creating a meditation garden involves intentional

design that invites tranquility.

Structural Elements:

Paths: Use materials like gravel, wood chips, or flat stones to guide movement.

Focal Points: Include a single, eye-catching feature like a small statue, a unique tree, or a natural sculpture.

Plant Selection:

Choose low-maintenance plants with soothing colors. Blue and green tones (think hostas, lamb's ear, or junipers) are naturally calming.

Include perennials like lavender or rosemary for continuity year after year.

Lighting:

Install solar-powered lanterns or string lights for evening meditation sessions.

Tip: Incorporate a spiral herb garden into your meditation space. The spiral shape naturally draws focus inward, promoting mindfulness while also being functional.

Outdoor Baths and Showers

An outdoor bathing space is an ultimate retreat feature.

DESIGN BASICS

Use salvaged materials like old galvanized tubs for rustic charm or repurpose wine barrels for a smaller soaking option.

Plant tall grasses, bamboo, or climbing vines on trellises around the bath for privacy.

Water Heating:

Solar heating: Position a black water tank nearby to absorb sunlight.

Wood-fired heaters: Build a small firebox with copper piping to circulate and heat water naturally.

Tip: Enhance the space with DIY bath products made from your homestead, such as oat and lavender bath sachets or honey-based face masks.

Herbal Treatments

HERBAL FACIAL STEAMS

Facial steams cleanse your skin and soothe your mind.

Recipe for Relaxation:

Combine dried chamomile, rose petals, and a sprig of rosemary in a bowl of hot water.

Add a few drops of essential oil (e.g., lavender for relaxation or eucalyptus for sinus relief).

Drape a towel over your head to create a tent, leaning over the bowl for 10–15 minutes.

Preserving Herbs:

Dehydrate herbs using a solar dehydrator to create a year-round supply. Store them in airtight jars.

ESSENTIAL OIL CRAFTING

Extracting essential oils is an advanced yet incredibly rewarding process.

Building an Outdoor Still for Essential Oil Extraction

Creating an outdoor still to craft essential oils is a rewarding homestead project that allows you to extract the aromatic and therapeutic properties of herbs. Below are detailed instructions for building an outdoor still and step-by-step guidance for using it.

MATERIALS NEEDED

For the Still

Copper or Stainless Steel Pot (5-gallon capacity): Serves as the boiling chamber.

Copper Tubing (20 feet, ¼-inch diameter): Used for the condenser coil.

Heat-Resistant Sealant (e.g., food-grade silicone): For sealing joints.

Large Glass Collection Jar: To capture the essential oil and hydrosol.

Metal Lid (to fit the pot): For creating a secure top for the still.

Food-Grade Steam Basket or Mesh: To hold the plant material.

Bucket or Tub (5-gallon): For the cooling water around the condenser coil.

Ice or Cold Water: Used for the cooling process.

Rubber Grommet or Bulkhead Fitting: To connect the tubing to the pot lid.

Clamps or Hose Clamps: To secure the tubing.

Tripod Stand or Bricks: To elevate the pot over a heat source.

Heat Source: A propane burner or campfire.

Tools Needed

Drill with a metal bit (½-inch diameter)
Pliers
Pipe cutter (for trimming copper tubing)
Wrenches
Hammer (for securing parts if necessary)
Funnel (optional for filling water)

CONSTRUCTION INSTRUCTIONS

1. Prepare the Boiling Chamber

Drill a hole in the lid of the pot, large enough to fit the rubber grommet or bulkhead fitting snugly.

Insert the copper tubing into the hole, ensuring it extends 3–4 inches into the pot. Seal with heat-resistant sealant to prevent leaks.

2. Build the Condenser Coil

Wrap the copper tubing around a cylindrical object (like a large coffee can) to form a coil. Ensure the coil is compact but allows water flow between the loops.

Place the coil inside the bucket or tub, leaving both ends of the tubing sticking out.

3. Assemble the Cooling System

Attach the top end of the coil to the tubing extending from the pot lid. Secure the connection with a clamp and seal with heat-resistant sealant.

Position the lower end of the coil to drain into the glass collection jar.

4. Secure the Pot

Elevate the pot on a tripod stand or bricks over the heat source, ensuring it's stable and level.

5. Test for Leaks

Fill the pot with water, seal the lid, and heat to check for steam escaping from any joints. Seal leaks as needed.

Using the Still

Step 1: Gather and Prepare Herbs

Collect fresh or dried herbs from your garden.

Examples include lavender, mint, rosemary, and chamomile.

Chop herbs coarsely to maximize surface area for steam contact.

Step 2: Load the Still

Place the herbs in the steam basket or mesh and position them inside the pot above the water level. Do not submerge them in water, as steam, not boiling, extracts essential oils.

Add enough water to the pot to produce steam without drying out during distillation (approximately 2–3 gallons).

Step 3: Start the Distillation

Light your heat source and bring the water to a gentle boil.

As the steam rises, it will pass through the plant material, capturing essential oils.

The steam travels through the tubing into the condenser coil, where it cools and condenses into liquid.

Step 4: Collect the Output

The liquid dripping from the end of the condenser coil will consist of essential oil (floating on top) and hydrosol (the water-based byproduct).

Use the glass jar to collect this mixture.

Step 5: Separate the Essential Oil

Allow the liquid to settle. The essential oil will rise to the top.

Use a pipette or separatory funnel to carefully extract the oil. Store it in dark glass bottles to protect it from light degradation.

TIPS FOR EFFECTIVE USE

Cooling System: Continuously replenish cold water or ice in the condenser bucket to maintain efficient condensation.

Monitor Heat: Avoid overheating, as this can burn the herbs and degrade the oil quality.

Herb Ratios: Use approximately 1–2 pounds of plant material per gallon of water for optimal yield.

BENEFITS OF AN OUTDOOR STILL

Cost-Effective: Homemade essential oils can save significant money compared to store-bought versions.

Eco-Friendly: Uses renewable herbs and materials.

Customizable: Allows you to craft blends tailored to your needs.

Self-Sufficiency: Provides a sustainable way to make therapeutic and aromatic products for personal use or sale.

By following these steps, you can craft a reliable and efficient still to produce high-quality essential oils from your homestead herbs.

Tip: Combine your homemade oils with beeswax and carrier oils (e.g., almond or jojoba) to make multipurpose balms.

Nature-Inspired Crafts

Use your environment as both material and inspiration.

Pressed Wildflower Art: Gather seasonal flowers and press them between heavy books. Create bookmarks, greeting cards, or framed art pieces.

Basket Weaving: Harvest flexible materials like

willow, honeysuckle, or ivy to craft durable and decorative baskets.

DYEING FABRICS

Foraging for dyes: Use black walnut for rich browns, goldenrod for yellows, and elderberries for purples.

Boil materials with a mordant like alum to fix the color.

CREATIVE WRITING AND JOURNALING

Turn your homestead journey into an artistic outlet.

Writing Prompts:

Describe the first signs of each season on your land.

Chronicle your personal growth alongside the growth of your homestead.

Publishing Potential: Consider turning your journal entries into a blog or book, sharing your unique insights with a wider audience.

Practical Tips for Implementation

BUDGET-FRIENDLY OPTIONS

Build retreat features with scavenged or second-hand materials. Old windows can create a rustic greenhouse; pallets can form seating or fencing.

Trade with neighbors for plants, materials, or labor.

ADAPTING FOR SMALL SPACES:

Even a balcony or tiny yard can house a mini Zen

garden or meditation nook. Use pots and vertical gardening techniques to maximize space.

Designing a homestead retreat is an ongoing journey of creativity and self-care. By integrating wellness practices, building peaceful spaces, and using your homestead's bounty for crafts and therapies, you create a holistic lifestyle that nourishes not just your body but also your soul. With a well-designed retreat, your homestead becomes more than a place of production—it becomes your sanctuary.

CHAPTER TWELVE: HOMESTEADING WITH TECHNOLOGY

Homesteading is a return to roots, but that doesn't mean we should turn our backs on the very tools that can help us achieve more while working less. In fact, integrating modern technology into homesteading can make your life simpler, more efficient, and more sustainable.

The goal is not to replace traditional methods but to enhance them. By strategically adopting tools, apps, and systems that align with your values of sustainability and self-sufficiency, you can optimize your time, reduce waste, and free yourself from the grind.

In this chapter, I touch on the smart tools that can give your homestead a competitive edge—whether you're growing food, managing livestock, or saving energy. These technologies are not just about convenience, they're about living more harmoniously with the land while making informed, data-driven decisions.

Farm Management Software

The modern homesteader needs to be a jack-of-all-trades, managing crops, livestock, financial records, and everything in between. Traditional methods of keeping track—jotting down planting dates, health logs, and inventory on paper or spreadsheets—can quickly become overwhelming. Farm management software takes the guesswork out of running a homestead, and it's not just for big farms.

Whether you have a small plot or a larger operation, using the right software allows you to track everything from soil conditions to income in one centralized location.

FARM MANAGEMENT APPS

Farmbrite – This user-friendly app allows you to create customized farm plans and keep track of everything from soil testing to planting schedules. It helps you organize your planting plans, harvest schedules, and crop rotations. It even helps track livestock health, providing detailed records of vaccinations, breeding, and medical treatments. Farmbrite is great for homesteaders who want to manage multiple aspects of their farm seamlessly and ensure productivity without the chaos.

Trello – Not strictly a farm management app, Trello's flexibility makes it an excellent tool for homesteading. It's essentially a digital bulletin board where you can create boards for specific projects (e.g., crop planting, animal care, homestead finances). Use it to break down big tasks into smaller steps, assign due dates, and track progress. You can set up checklists for your weekly tasks, whether that's feeding the animals, harvesting crops, or setting up your irrigation system.

AgriWebb – This software is ideal for livestock management. It lets you track animal health, movements, breeding cycles, and pastures. You can also use it to schedule and record feeding, vaccinations, and hoof care. The added benefit? You can use AgriWebb to track the financial side of raising livestock, ensuring that you can optimize your income while managing expenses.

Tip:

Use **FarmLogs** alongside your farm management software. FarmLogs specializes in tracking weather conditions, soil health, and crop yield predictions.

By syncing it with a software like Farmbrite, you get a holistic view of your homestead's performance.

The weather data integration allows you to predict harvest windows and track how weather conditions (like droughts or excessive rain) are affecting your crops.

Smart Irrigation Systems

Watering your crops and garden is a critical component of homesteading, and managing it efficiently can significantly reduce your water usage while ensuring your plants stay hydrated.

Traditional irrigation systems often waste water through evaporation or runoff, but a smart irrigation system automates the process with precision, saving both water and time.

TYPES OF SMART IRRIGATION SYSTEMS:

Drip Irrigation Systems: Drip irrigation is one of the most efficient ways to water your garden because it delivers water directly to the plant roots, minimizing evaporation and runoff. **Netafim** offers advanced drip systems that come with smart controllers to regulate water flow based on moisture levels in the soil. By installing a system that integrates with a moisture sensor, you can reduce water usage while keeping your crops hydrated.

Install a timer or smart controller like the **Rachio 3**, which can sync with local weather forecasts to automatically adjust your watering schedule. If a storm is expected, it will skip the watering session to conserve water, ensuring you don't water right before rainfall.

Soil Moisture Sensors: Add soil moisture sensors like **The Vegetronix VH400** or **HydroPoint** to your garden to accurately measure the moisture content at different depths of your soil. These sensors send real-time data to your irrigation system, triggering watering only when necessary. This precise, on-demand watering keeps your plants happy and saves you from overwatering—important when drought conditions are a concern.

Smart Sprinkler Systems: If your homestead spans several acres or you need to irrigate large, open spaces, a smart sprinkler system like **Rachio Smart Sprinklers** or **Rain Bird** can manage multiple zones automatically. These systems allow you to program different watering schedules for areas with diverse needs, like a vegetable garden versus a lawn or orchard.

For homesteaders who need to irrigate using well water, installing a solar-powered water pump can significantly reduce your reliance on the grid. Systems like the **SHURflo 9300** can move water from your well to your crops using solar power, reducing operating costs and providing reliable water access without drawing from your electrical supply.

Solar-Powered Gadgets

Solar power isn't just for powering your house—it's for powering your homestead's tools, gadgets, and even systems you rely on every day. Solar-powered devices not only reduce your environmental footprint but can be incredibly cost-effective over time.

SOLAR-POWERED TOOLS

Solar Water Pumps: Solar water pumps are especially useful for homesteaders who rely on wells or ponds to irrigate their crops. These pumps are powered by solar panels and can be used to move water from a well, creek, or rainwater harvesting system into a storage tank or irrigation line. **SHURflo** and **Grundfos** are top brands offering reliable solar-powered pumps. These systems can be set up for everything from livestock watering to garden irrigation.

If you have a large garden or orchard, consider installing a solar-powered irrigation timer to regulate when and how much water is delivered to your plants. Systems like **HydroPoint's smart controllers** are perfect for monitoring and optimizing irrigation based on solar power.

Solar-Powered Dehydrators: Solar dehydrators are perfect for preserving the food grown on your homestead without using grid power. These dehydrators use solar energy to gently remove moisture from fruits, vegetables, and herbs.

Sun Ovens and **All-American Sun Oven** make solar dehydrators that can preserve everything from tomatoes to herbs, helping you stock your pantry with dried food for the winter.

Solar-Powered Lighting: Whether you're lighting up a path or providing illumination to your barn or greenhouse, solar-powered lights are an excellent way to reduce electricity costs and keep your homestead illuminated after dark. Companies like Gama Sonic and Moonrays offer solar-powered lights that are easy to install, with no wiring required.

To maximize solar power use, consider investing in a solar power bank or solar inverter. These devices can store excess solar energy for use later in the day, ensuring you have power available for critical systems (like refrigeration or water pumps) when the sun isn't shining.

Monitoring Systems

As a modern homesteader, your success is often determined by the data you collect and how you act on it. Monitoring tools provide real-time insights that help you make informed decisions about everything from soil health to weather patterns, livestock management, and crop conditions.

KEY MONITORING TOOLS

Weather Stations: A local weather station, like Netatmo Weather Station or Ambient Weather, is invaluable for real-time updates on temperature, humidity, wind speed, and barometric pressure. By understanding local weather conditions, you can make critical decisions, such as when to plant crops, when to harvest, or when to take protective measures against extreme weather events.

Some weather stations, like WeatherFlow, can integrate with apps on your phone, providing hourly and daily forecasts. This allows you to adjust your homestead activities accordingly—whether that's protecting your garden from an impending frost or taking advantage of a sunny day to power your solar systems.

Soil Sensors: For a more hands-on approach, soil sensors like The Vegetronix VH400 or **Sentek's Drill-and-Drop** sensors monitor moisture levels and pH, helping you fine-tune your irrigation and fertilization schedules. These sensors allow you to maintain the ideal conditions for your crops, ensuring optimal growth.

Livestock Health Monitoring: For larger homesteads with animals, health monitoring systems like CowManager or Fitbit for Cows provide valuable data about livestock activity, fertility, and health. These devices are especially useful for monitoring the reproductive cycles of animals or tracking health metrics for disease prevention.

Consider integrating all your monitoring systems (weather, soil, livestock) into a centralized dashboard using platforms like FarmOS or SmartFarm. This allows you to see a live, holistic overview of all critical data, and make real-time decisions from anywhere on your property.

TIPS

Start Small and Scale: Don't overwhelm yourself with every tool available. Begin with one or two systems that make the biggest impact on your homestead. Once you're comfortable, expand gradually. For example, start with a weather station and a drip irrigation system, and build from there.

Use Data Wisely: With all the data at your fingertips, the real challenge is interpreting and using it effectively. Make sure you regularly review your monitoring systems to make informed adjustments to your practices.

Technology is an invaluable tool that can help homesteaders manage resources, improve productivity, and reduce costs. By using the right mix of traditional methods and modern tools, you can achieve a level of efficiency and sustainability that previous generations only dreamed of. However, the true power lies in your ability to adapt and integrate these tools into a lifestyle that values self-sufficiency while maintaining a deep connection with the land. The key is to balance innovation with intention, using technology to enhance your homesteading experience rather than overshadow it.

CHAPTER THIRTEEN: INCOME GENERATION

Let's talk about moving beyond self-sufficiency to creating a truly self-sustaining income stream from your homestead. In this chapter, we're going to explore how to transition from merely growing and producing for personal use to creating a profitable business model that can provide both financial freedom and the satisfaction of offering high-quality, handmade products to your community or even global markets.

Selling Produce, Eggs, Dairy, and Crafts: Scaling Your Homestead Output

Your journey into homesteading has already established the fundamentals of growing food and raising livestock. Now, let's explore how to turn these into profitable, scalable ventures.

SELLING FRESH PRODUCE: EXPANDING FOR PROFIT

One of the most natural and rewarding ways to generate income is by selling the food you've already cultivated with care. But scaling up requires a strategic approach. Here's how you can take your gardening to the next level and turn it into a sustainable business:

Advanced gardeners know the value of planning for continuous harvests. For example, planting two or even three rounds of crops in one season (succession planting) can increase your output and market availability.

Rather than letting your beds sit idle, plan for rotations. For example, planting early-season lettuce in early spring, followed by radishes or kale, then late-season vegetables like broccoli, means you'll have fresh produce through the entire growing period.

MICROGREENS & HIGH-DEMAND CROPS

Consider growing microgreens, which have a fast turnover and are highly profitable. Microgreens like arugula, radishes, and mustard greens can be harvested in as little as 10–14 days, making them an ideal crop for farmers' markets or subscription boxes.

Specialty Crops for Higher Margins: While tomatoes and potatoes are reliable, consider diversifying with high-margin crops like heirloom tomatoes, gourmet garlic, or edible flowers. You can even look into mushrooms—like shiitakes or oyster mushrooms—that require little space and offer great returns. Specialty crops not only sell for higher prices but also create a niche market.

Marketing and Selling Produce:

Subscription Boxes & CSA (Community Supported Agriculture): Establishing a CSA program where customers subscribe to receive weekly or bi-weekly boxes of seasonal produce is an excellent way to guarantee sales before harvest. Many homesteaders find success with this model, especially if they market their unique, locally-grown goods. Websites like Farmigo can help set up CSA systems, manage subscriptions, and automate some of the logistics.

Online Sales Platforms: Consider selling via platforms like LocalHarvest or creating your own Shopify store. You can also partner with local organic food delivery services. These methods allow you to access customers beyond your immediate area.

SELLING EGGS AND DAIRY

Eggs and dairy products often have a premium price, particularly when marketed with an emphasis on quality and care. To maximize profits, explore ways to differentiate yourself in the marketplace.

MAXIMIZING EGG AND DAIRY SALES

Rare or Heritage Eggs: Specialize in eggs from rare or colorful breeds, such as Marans (dark brown eggs) or Ameraucanas (blue eggs). These eggs not only have a unique selling point but can also command much higher prices.

Dairy Products: If you're producing milk from goats or cows, transform this milk into high-value products such as cheese, yogurt, kefir, or even skincare products like goat milk soap. For example, artisan cheeses like feta or chèvre can be sold for $10–20 per pound depending on the quality and rarity. But always check your local health regulations for dairy production licenses.

Raw Milk & Specialty Dairy Goods: Raw milk, although regulated in some states, can be an incredibly lucrative product if you have the legal right to sell it. Offering flavored milk (e.g., chocolate or strawberry) or kefir can cater to health-conscious consumers looking for probiotic-rich foods.

SELLING STRATEGIES

Farmers Markets: These remain one of the best ways to sell eggs and dairy directly to consumers, especially if you offer organic or heirloom breeds. Set up a visually appealing booth, complete with a product list and a story behind your farming practices. Building relationships at these markets can lead to repeat customers.

Online Platforms: Platforms like Etsy are also a great place to market value-added dairy products. Additionally, you could consider creating a subscription box of seasonal dairy items or offering products to local grocery stores or restaurants.

Crafting for Profit

Homesteading provides the perfect foundation for crafting goods that can be sold at local markets or online. Whether you're making soap, candles, pottery, or woodworking items, your creativity combined with sustainable materials gives you a competitive edge.

CRAFTING IDEAS FOR HOMESTEADERS

Soap Making with Homestead Ingredients: Homemade soap made from goat's milk or infused with herbs like lavender, mint, or chamomile is both a profitable and highly sought-after product. Start by offering small-batch artisanal soap bars with eco-friendly packaging.

Handcrafted Pottery or Wooden Goods: Selling handmade pottery (e.g., mugs, planters, etc.) or wooden items like spoons, chopping boards, or furniture allows you to create a niche market for rustic, handcrafted, high-quality products.

SELLING YOUR CRAFTS

Etsy & Local Art Fairs: Etsy is the go-to platform for homesteaders looking to sell their handmade goods. With an emphasis on artisanal, organic, and sustainable products, it provides access to a global audience. In addition, participating in local art fairs or hosting workshops can allow you to build local recognition and loyal customers.

Wholesale Opportunities: If you create crafts at scale, reaching out to small boutique stores or gift shops to offer wholesale deals can expand your reach. Small homestead-themed gift shops may be especially receptive to stocking your goods.

Hosting Workshops or Offering Tours

One of the best ways to scale your homestead business is by sharing the knowledge you've spent years cultivating. Hosting workshops or offering tours can generate significant income, especially if you can capitalize on the growing interest in self-sufficiency and sustainable living.

WORKSHOP IDEAS

Canning and Preserving: Canning is a perennial favorite for homesteaders, and teaching workshops on this can bring in participants eager to learn how to preserve food for the future. You can teach everything from pickling vegetables to pressure canning meats.

DIY Projects and Building: Offering workshops on building essential homestead tools, such as raised beds, chicken coops, or composting systems, not only helps others but also provides a solid income stream.

TOURS

Permaculture & Homesteading Tours: If you've implemented permaculture techniques, offer guided tours that demonstrate water conservation, soil regeneration, and the creation of closed-loop ecosystems. Visitors will appreciate learning how to apply these principles to their own land.

Lifestyle Tours: Some homesteaders are able to provide immersive experiences where guests can observe or participate in the daily activities of farm life—gathering eggs, feeding animals, and more. You can charge a fee for these tours and offer products for sale as part of the experience.

Selling Value-Added Products

Value-added products are a fantastic way to increase the value of your crops and livestock. These are items that involve more work than simply selling raw products but can be highly profitable.

EXAMPLES OF VALUE-ADDED PRODUCTS

Preserves, Jams, and Jellies: Homemade preserves made from fruit and herbs grown on your property are highly profitable, especially when you add unique flavors, like lavender jelly or habanero peach preserves. These can be sold at farmers' markets or packaged in beautiful jars for online sales.

Herbal Salves and Tinctures: If you're growing medicinal herbs, creating herbal tinctures, balms, or oils can be a lucrative business. Start by focusing on a niche, such as stress-relief or skincare, and expand your offerings as demand grows.

Online Business Opportunities

The digital age has made it easier than ever to market your homesteading skills or products. From blogging to social media and online stores, homesteaders have access to global markets.

BLOGGING & YOUTUBE

Start a Blog: Share your homesteading journey through blogging. Topics might include garden tips, livestock care, DIY projects, or sustainable living practices. Monetize your blog through ads, sponsored content, or affiliate marketing.

YouTube Channel: For those who are comfortable

on camera, creating a YouTube channel dedicated to homesteading can also be an excellent way to generate income. Tutorials, product reviews, and vlogs can all attract audiences eager to learn about sustainable living.

By strategically focusing on high-value, niche products, workshops, and online platforms, you can transform your homestead into a sustainable and profitable venture. The key to success in homestead businesses is understanding your strengths and leveraging your unique products and skills to create a niche market. Step by step, you're not only growing food and raising animals; you're growing the business of your dreams.

CHAPTER FOURTEEN: LOOKING FORWARD

Congratulations! You've invested countless hours and resources to get to where you are today. You've built a homestead from the ground up, learned the foundations, and cultivated your own small corner of self-sufficiency.

But here's the truth: becoming fully self-sufficient is a journey, not a destination. Let's move from just surviving to thriving, from the basics to advanced techniques in food production, energy independence, and sustainability. Your homestead can be a resilient and thriving force, capable of adapting to challenges and growing stronger with each season.

It's time to embrace new skills, strategies, and methods to truly master the art of self-sufficiency.

Continuous Learning and Adapting

As much as you've learned and accomplished so far, homesteading is an evolving practice. You can never truly "know it all" because the world of self-sufficiency is full of endless learning opportunities. This mindset will be your greatest asset in ensuring your homestead thrives.

EXPERIMENTATION

The key to true homesteading mastery is your ability to experiment with new ideas and technologies. Don't be afraid to test out different practices, even if they seem a little outside the box. For example, consider experimenting with crop rotation in a way that you haven't tried before.

Maybe you can grow nitrogen-fixing legumes one year and deep-rooted root crops the next. You might find that a certain variety of squash grows better in your soil than the one you've always used.

If you're growing perennials, try planting some edible groundcovers like creeping thyme or strawberries, which can provide food while enriching the soil for other crops. Experimenting with different techniques can bring unexpected results, such as reduced pest infestations or even a larger harvest with less work.

COMMUNITY LEARNING

You don't have to reinvent the wheel—reach out to the global homesteading community. Whether it's through online forums or local meet-ups, there's a wealth of knowledge waiting for you. One tip from a seasoned gardener might lead to a breakthrough in your crop yields. A friend who raises pigs may have valuable insights on rotational grazing that will save you time and effort. Homesteading isn't just about what you know—it's about who you can learn from, too. Seek out mentors, attend workshops, or even host your own.

EMBRACE TECHNOLOGICAL ADVANCES

Just because you hope to live off the grid doesn't mean you can't embrace some modern technology. In fact, using tools like weather apps or automated irrigation systems can save you tons of time and effort while maximizing the efficiency of your homestead. Take a look at solar-powered water pumps, automatic greenhouse ventilation systems, or even weather stations that help you predict the best times to plant or harvest. These tools are often much simpler to implement than you might think—and they can take your homesteading operation to the next level.

Remember: as long as you stay curious and open to change, your homestead will continue to grow and evolve, just like you.

Planning for Contingencies

Every homesteader will encounter challenges—weather extremes, pest invasions, or even animal illness. The difference between thriving and surviving often boils down to how well you plan for the unexpected.

DROUGHT & EXTREME WEATHER

Water is the lifeblood of your homestead, and in times of drought or extreme heat, it can become the most precious resource. Advanced water management techniques are non-negotiable.

Rainwater Harvesting: Start by building a system to collect and store rainwater. A simple setup can include gutters on your barn or house, a storage tank, and a filtering system to ensure clean water for your plants and animals. For larger systems, you might consider a series of interconnected rain barrels or even a custom-built cistern.

Greywater Recycling: If you haven't already, consider setting up a greywater recycling system. This takes water from sinks, showers, or laundry (with eco-friendly detergents) and redirects it to your garden. Not only will this save water, but it will also provide your plants with nutrient-rich runoff.

Drought-Resistant Crops: When selecting crops, always opt for drought-resistant varieties. Crops like sorghum, quinoa, millet, and buckwheat are drought-tolerant and can thrive in tough conditions. For vegetables, look at varieties like okra, sweet potatoes, and zucchini, which require less water. Keep in mind, the water use of each plant varies, and careful selection can make a huge difference in dry seasons.

CROP FAILURE

Not every season will be perfect, and crop failure

can happen. But if you diversify your crops, you'll have a built-in insurance policy against total failure.

Diversify & Rotate: By planting a variety of crops that mature at different times, you spread the risk. Try not to put all your resources into a single crop. Instead, grow a mix of annuals and perennials, such as beans, tomatoes, carrots, and leafy greens. If one crop fails, another may thrive.

Soil Health: Focus on building long-term soil health. Use compost, mulch, and cover crops to improve your soil structure and nutrient content. You might consider using biochar or rock phosphate to enhance soil resilience, especially during dry years. Building a healthy, nutrient-dense soil ecosystem will make your plants more resistant to disease, drought, and pests.

ANIMAL ILLNESS

The health of your animals is just as critical as the health of your crops. Keep them disease-free and stress-free, and you'll minimize the risk of costly or fatal illnesses.

Preventative Care: Ensure that your animals are vaccinated and dewormed regularly. Check them for signs of illness, such as changes in behavior, appetite, or the appearance of their skin and fur. Always keep track of vaccination schedules and seek advice from a local vet when necessary.

Emergency Health Kits: Prepare an emergency health kit for each type of livestock you own. This should include essential supplies such as bandages, disinfectants, electrolytes, antibiotics (where allowed), and any specific medications your animals may need.

Soil Regeneration and Seed Saving

As your homestead grows, the most important

resource to preserve is your soil. Without healthy soil, your crops won't thrive, and without reliable seed sources, you may find yourself reliant on external markets. Long-term sustainability is built on solid soil practices and a commitment to preserving your seeds.

SOIL REGENERATION

Advanced soil health practices focus on building up, rather than depleting, your land's fertility. One of the most effective ways to do this is by implementing regenerative practices.

Vermiculture & Composting: Go beyond basic composting and implement vermiculture (worm farming) to add beneficial microbes to your soil. Worm castings are some of the most nutrient-dense fertilizers available and can improve your soil's structure. Create compost bins that include not only kitchen scraps but also animal manure and wood chips. Hot composting can break down organic matter in just a few months, enriching the soil with valuable organic material.

Cover Crops & Crop Rotation: Use cover crops like clover, rye, and vetch during the off-season to restore nitrogen levels in the soil and prevent erosion. Crop rotation is critical for reducing soil depletion and pest buildup. After each season, change where your plants grow to reduce disease and pest infestations.

SEED SAVING

Seed saving is the foundation of long-term self-sufficiency. You can ensure that you're always able to grow the food you need without relying on external seed sources.

Collecting Heirloom Seeds: Choose heirloom, open-pollinated varieties that will reliably reproduce year after year. Some of the best seeds to save include tomatoes, beans, squash, and lettuce. Make sure to harvest seeds at the right time—when they're fully mature but before they're exposed to pests or disease.

Storage & Preservation: Store your seeds in airtight containers in a cool, dry, and dark place. Glass jars with tight-fitting lids or Mylar bags work well. If you've saved a lot of seed, you can freeze them to extend their viability, especially for long-term storage.

Scaling Up for Larger Homesteads

Now that you've built a solid foundation, it's time to think about scaling up. But scaling doesn't mean just buying more land or animals; it's about expanding efficiently and sustainably.

EFFICIENT LAND USE

Zoning your homestead is crucial for maximizing productivity. Identify areas for growing high-demand crops, areas for perennial plants, livestock pens, and other important functions. Use techniques like agroforestry, where trees, shrubs, and crops work together to maximize space and soil fertility. Integrate perennial plants like fruit trees and berry bushes into your landscape to reduce maintenance time.

BUSINESS SCALING

If you're interested in generating income, think beyond just produce. Consider offering value-added products like homemade jams, soaps, or even wool products if you raise sheep.

You can also sell online, expand your presence at local farmer's markets, or start your own farm store.

Business scaling means taking what you've learned and turning it into a sustainable income stream.

Feel free to check out the other books in my Homesteading series which go even deeper into concepts and ideas touched on in this book. Together, we'll keep learning, growing, and thriving—one season at a time.

Becoming fully self-sufficient is a lifelong journey. It's about embracing new challenges, learning from each experience, and adapting your practices to the changing world around you. Whether you're managing water resources, caring for livestock, or preparing for the unexpected, remember that your success is built on persistence, flexibility, and continuous growth.

As you move forward, keep your focus on long-term sustainability. Build healthy soil, save your seeds, and embrace new ideas. The journey to full self-sufficiency may be challenging at times, but the rewards are immeasurable: a homestead that thrives, produces, and sustains you and your family for years to come.

Stay curious, keep learning, and always push forward. You've got this!

BOOKS IN THIS SERIES

The Homestead Blueprint: Introductory Self-Sufficiency With Homesteading

Homesteading Mastery: Advanced Skills and Techniques on the Modern Homestead

Homesteading for Total Self-Sufficiency

The Urban and Suburban Homestead

Homesteading and Natural Building Techniques

Sustainable Homestead Design and Permaculture Principles

Finding Your Homestead: A Comprehensive Guide to Purchasing Land in the U.S., Canada, and Europe

FURTHER RESOURCES FOR SUCCESS

Books

Animal, Vegetable, Miracle: A Year of Food Life by Barbara Kingsolver: A memoir that explores sustainable living through local food production and seasonal eating.

Back to Basics: A Complete Guide to Traditional Skills by Abigail Gehring: This book offers a wealth of information on a wide range of homesteading skills.

*Beekeeping for Dummie*s by Howland Blackiston: A comprehensive guide covering the basics of beekeeping, from setting up hives to harvesting honey.

Building Soils for Better Crops by Fred Magdoff and Harold van Es: A guide on organic soil management and improvement.

Canning and Preserving for Dummies: A user-friendly introduction to canning, with plenty of practical tips and recipes.

Carrots Love Tomatoes: Secrets of Companion Planting for Successful Gardening by Louise Riotte: This book provides a wealth of information on companion planting, including specific pairings and gardening tips.

DIY Hydroponic Gardens by Tyler Baras: Provides step-by-step instructions for setting up hydroponic systems at home.

DIY Solar Projects by Eric Smith: A practical guide to implementing solar energy systems at home.

Duck Eggs Daily by Lisa Steele: A comprehensive guide to raising ducks.

Ferment Your Vegetables by Kirsten K. Shockey and Christopher Shockey: A user-friendly book with a variety of recipes and techniques for fermenting vegetables.

Four-season Harvest by Eliot Coleman: This book is a fantastic resource for extending your gardening season and includes detailed information on cold-hardy crops and techniques.

Gaia's Garden: A Guide to Home-Scale Permaculture by Toby Hemenway: Provides insights into designing sustainable garden systems.

Let It Rot!: The Gardener's Guide to Composting by Stu Campbell: A classic guide that covers the ins and outs of composting.

Permaculture: A Designer's Manual by Bill Mollison

Preserving Food Without Freezing or Canning by The Gardeners and Farmers of Terre Vivante: Explores alternative methods of food preservation that don't involve canning or freezing.

Rainwater Harvesting for Drylands and Beyond by Brad Lancaster: A detailed resource on harvesting rainwater and using it effectively.

Raised Bed Gardening for Beginners by Annie Schreiber: A straightforward book that covers everything from building beds to planting and maintaining them.

Raising Rabbits for Meat by Julie S. Baker: A practical guide for those focused on meat production, including tips on care and processing.

Resilience: Hard-Won Wisdom for Living a Better Life by Eric Greitens: A profound exploration of resilience and how to cultivate it in life.

Root Cellaring: Natural Cold Storage of Fruits & Vegetables by Mike Bubel and Nancy Bubel: A guide to preserving produce through root cellaring techniques.

Seed to Seed by Suzanne Ashworth: A detailed guide on saving seeds from a variety of plants, ideal for sustainable gardening.

Solar Power for Dummies by Rik DeGunther: An excellent guide to understanding solar energy systems.

Square Foot Gardening by Mel Bartholomew: A classic guide on this technique.

Storey's Guide to Raising Chickens by Gail Damerow: A comprehensive guide on everything from chicken breeds to coop building and care.

Storey's Guide to Raising Goats by Margaret Hathaway: Offers insights into goat care, breeding, and production.

Storey's Guide to Raising Rabbits by Bob Bennett: This comprehensive guide covers everything from housing to breeding and health care.

Teaming with Microbes by Jeff Lowenfels and Wayne Lewis: A comprehensive guide to understanding soil health and the role of microbes.

The Art of Fermentation by Sandor Katz: A detailed exploration of fermentation techniques for preserving food and improving health.

The Backyard Homestead: Produce All the Food You Need on Just a Quarter Acre! by Carleen Madigan. This book covers a broad range of topics for backyard farming.

The Backyard Homesteader's Book of Kitchen Wisdom by Anne D. Kinsinger: Focuses on transforming homegrown produce into delicious meals and preserves.

The Ball Complete Book of Home Preserving: This comprehensive guide covers everything from canning basics to advanced techniques.

The Beekeeper's Bible by Richard A. Jones and Sharon Sweeney-Lynch: A thorough guide covering everything about beekeeping, including history, tools, and techniques.

The Chicken Encyclopedia by Gail Damerow: A go-to reference for chicken keepers, offering detailed information on care, breeds, and behavior.

The Chicken Whisperer's Guide to Keeping Chickens by Andy Schneider: A great book for beginners with practical tips and solutions to common problems.

The Complete Book of Home Preserving by Judi Kingry

The Complete Book of Sewing by Alison Smith: A thorough resource for mastering sewing techniques, from basic to advanced.

The Complete Compost Gardening Guide by Barbara Pleasant and Deborah L. Martin: Offers practical advice on composting techniques and troubleshooting.

The Complete Dehydrator Cookbook by Janet Hurst: A comprehensive guide with recipes and techniques for dehydrating a wide range of foods.

T*he Complete Guide to Edible Wild Plants* by the U.S. Department of Agriculture: An essential guide for foraging and utilizing wild edibles.

The Complete Guide to Raising Rabbits by Kim McGowan: A comprehensive resource covering every aspect of rabbit care, from breeds to breeding.

The Complete Guide to Sharpening by Leonard Lee: A fantastic resource for learning how to sharpen all types of tools, from pruners to axes.

The Complete Manual of Woodworking by Albert Jackson and David Day: Covers everything from basic carpentry skills to advanced woodworking techniques

The Compost Gardener's Handbook by Rhonda Massingham Hart: This book offers detailed insights into various composting methods and troubleshooting tips

The Dehydrator Bible by Amy Bronee, and others: Offers step-by-step instructions and creative recipes for dehydrated foods

The Encyclopedia of Country Living by Carla Emery: A comprehensive reference for homesteaders, covering a wide range of topics from gardening to livestock care.

The Fruit Gardener's Bible by Lewis Hill: Provides comprehensive information on growing and caring for fruit trees and bushes

The Garden Primer by Barbara Damrosch: An excellent resource that covers all aspects of vegetable gardening.

The Herbal Kitchen: Cooking with Spices, Herbs, and Edible Flowers by Jerry Traunfeld: Explores culinary uses for various herbs and spices, enhancing homesteading cooking skills.

The Herbal Medicine-Maker's Handbook by James Green

The Holistic Orchard: Tree Fruits and Berries the Biological Way by Michael Phillips: Focuses on sustainable fruit cultivation and orchard management.

The Lean Farm: How to Minimize Waste, Increase Efficiency, and Maximize Value and Profits with Less Work by Ben Hartman: This resource focuses on efficient farming practices that cultivate problem-solving mindsets.

The MeatEater Fish and Game Cookbook by Steven Rinella: A comprehensive cookbook for those who hunt or fish, focusing on preparation and cooking methods.

The Modern Survival Manual: Surviving the Economic Collapse by Fernando Aguirre: Offers practical strategies for self-sufficiency and resilience in challenging times.

The New Organic Grower by Eliot Coleman: This book provides invaluable insight into organic gardening, including raised bed techniques.

The New Seed-Starter's Handbook by Nancy Bubel: Provides methods for starting seeds indoors, including tips for various types of plants.

The Organic Gardener's Handbook of Natural Pest and Disease Control by Fern Marshall Bradley: Focuses on natural methods to manage pests and diseases in the garden.

The Permaculture City by Toby Hemenway: Provides insight into food forest design and implementation.

The Permaculture Handbook: Garden Farming for Town and Country by Peter Bane: This book includes strategies for water management in permaculture settings.

The Permaculture Student by Andrew Millison: A resource for applying permaculture principles to create sustainable systems at home.

The Rabbit Raising Problem Solver by Karen Patry: This book provides solutions to common issues faced by rabbit owners, making it an invaluable resource.

The Renewable Energy Handbook by William H. Kemp: Provides practical advice on implementing renewable energy solutions.

The Resilient Farm and Homestead: An Innovative Permaculture and Whole Systems Design Approach by Ben Falk: This book covers holistic approaches to sustainable farming and problem-solving.

The Resilient Gardener: Food Production and Self-Reliance in Uncertain Times by Eliza Greenman: A practical guide to gardening with an emphasis on resilience and adaptability.

The Self-Sufficient Life and How to Live It by John Seymour: A classic guide that covers everything from gardening to livestock to alternative energy sources.

The Soil Book: How to Manage Your Soil for Productive Gardens by David R. Montgomery: Provides insight into soil health and practical amendment strategies.

The Soil Will Save Us by Kristin Ohlson: An exploration of how healthy soil can combat climate change and improve food security.

The Urban Homestead: Your Guide to Self-Sufficient Living in the Heart of the City by Kelly Coyne and Erik Knutzen: Ideal for those who want to homestead in small, urban spaces.

The Vegetable Gardener's Bible by Edward C. Smith: Provides detailed planting calendars and crop-specific information.

The Vegetable Gardener's Container Bible by Edward C. Smith: Offers a wealth of information on growing vegetables in tight spaces.

The Water-Wise Home by Laura Allen: A comprehensive guide to sustainable water management practices for homes and gardens.

The Winter Harvest Handbook By Eliot Coleman: Focused specifically on winter gardening, this book provides insights into greenhouse and cold frame

Urban Rooftop Farming by Aimee Rachel Plant: An in-depth look at turning rooftops into productive spaces.

Vertical Gardening: 100 Ways to Grow Up in a Down-to-Earth World by Derek Fell: Offers creative ideas and designs for vertical gardening.

Websites

American Rainwater Catchment Systems Association (ARCSA) (arcsa.org): Offers resources and educational materials on rainwater harvesting practices and regulations.

Backyard Chickens (backyardchickens.com): have guides on urban chicken keeping and city-specific regulations

Backyard Rabbits (backyardchickens.com/rabbits): A community-driven forum with discussions and advice on rabbit raising

Edible Landscapes (ediblelandscapes.com): A resource for creating beautiful, productive edible landscapes.

Farm & Garden (farmandgarden.com): Provides information on sustainable farming techniques, gardening, and livestock care.

Fermentation Therapy (fermentationtherapy.com): A resource for fermentation techniques and recipes for various foods and beverages.

Food Revolution Network (foodrevolution.org): Advocates for sustainable food systems and offers resources for healthy eating and cooking.

Garden Design (gardendesign.com): Features articles on garden planning, design, and sustainable landscaping.

Garden.org (garden.org): Provides plant care advice, gardening tips, and a community forum for gardeners.

Gardenista (gardenista.com): Offers inspiration and advice on gardening and outdoor living, with a focus on sustainable practices.

Good Housekeeping (goodhousekeeping.com): Features articles on home maintenance, gardening, and sustainable living practices.

Green America (greenamerica.org): Provides resources for ethical and sustainable living, including homesteading practices.

Heritage Farmstead Museum (heritagefarmstead.org): Offers educational resources and workshops on traditional farming and homesteading skills.

Homegrown Evolution (homegrownevolution.com): A site focused on DIY projects, urban homesteading, and sustainable living practices.

Homestead Honey (homesteadhoney.com): A blog dedicated to natural living, beekeeping, and homesteading practices.

Local Harvest (localharvest.org): Connects consumers with local farmers and markets for fresh produce and sustainable products.

Mother Earth News (motherearthnews.com): A treasure trove of articles on everything from organic gardening to DIY home projects.

Mother Nature Network (mnn.com): A site focused on environmental news and eco-friendly living, including gardening and homesteading tips.

National Gardening Association (garden.org): Offers extensive articles on growing fruit and perennial plants

National Institute of Food and Agriculture (nifa.usda.gov): Links to local extension offices

Permaculture News (permaculturenews.org): A valuable resource for innovative ideas and problem-solving in sustainable living.

Permaculture Research Institute (permaculturenews.org): A valuable source for permaculture design principles, including water management.

Rabbit.org (rabbit.org): A rich resource offering information on care, breeds, and health issues.

Rodale Institute (rodaleinstitute.org): A leading organization in organic farming, they provide research and resources on building healthy soil

Sew Mama Sew (sewmamasew.com): A resource for sewing patterns, tutorials, and a vibrant sewing community.

Square Foot Gardening Foundation (squarefootgardening.org): This site provides resources and techniques for growing food in small spaces using the square foot gardening method.

Survival Podcast (thesurvivalpodcast.com): Offers a wealth of information on survival skills, self-sufficiency, and homesteading.

Sustainable Agriculture Research and Education (SARE) (sare.org): This organization offers a wealth of information about sustainable soil practices and crop rotation

The American Rabbit Breeders Association (ARBA) (arba.net): A great resource for breed information, care guides, and local shows.

The Backyard Homestead (backyardhomestead.com): Offers a wide range of resources on growing food and raising animals in small spaces.

The Composting Council (compostingcouncil.org): Provides resources on composting techniques and benefits

The Good Trade (www.thegoodtrade.com): A

platform focused on sustainable living, ethical fashion, and conscious consumerism.

The Herbal Academy (theherbalacademy.com): Offers courses and resources on herbalism, plant medicine, and natural living.

The Homebrew Academy (homebrewakademy.com): A resource for home brewing beer and fermentation techniques.

The Homestead Survival (thehomesteadsurvival.com): Focuses on survival tips, homesteading how-tos, and DIY projects.

The Homesteading Hippy (homesteadinghippy.com): A blog offering practical tips and advice for various homesteading skills and mindsets.

The Horticult (thehorticult.com): A resource for urban gardening, DIY projects, and modern homesteading in small spaces.

The Kitchen Garden (kitchengardeners.org): Focuses on growing your own food and includes community resources and projects.

The National Center for Appropriate Technology (NCAT) (ncat.org): Offers resources and assistance for sustainable agriculture and energy practices.

The Old Farmer's Almanac (almanac.com): Offers planting calendars and gardening tips tailored to your region.

The Poultry Site (thepoultrysite.com): Offers a wealth of articles on chicken breeds, care, and health.

The Prairie Homestead (theprairiehomestead.com): A blog offering practical tips, recipes, and resources for modern homesteaders.

The Spruce (thespruce.com): Features articles on gardening, DIY projects, and sustainable living.

The Survivalist Blog (thesurvivalistblog.net): Offers resources and articles on preparedness, homesteading, and self-sufficiency.

The Xerces Society (xerces.org): Focuses on promoting sustainable landscaping and water conservation practices.

TreeHugger (treehugger.com): An online sustainability resource that covers eco-friendly living, gardening, and environmental news.

University of California Agriculture and Natural Resources (ucanr.edu): Offers extensive resources on irrigation practices and water management strategies

Urban Farming (urbanfarming.org): A platform focused on promoting urban agriculture and offering resources for urban homesteaders.

USDA Natural Resources Conservation Service (nrcs.usda.gov): Offers guidance on water conservation practices for gardens and landscapes.

Zero Waste Home (zerowastehome.com): A resourceful site offering tips and ideas for living a zero-waste lifestyle.

YouTube Channels

Annie's Kitchen Garden: Features tutorials on growing vegetables, canning, and cooking from the garden.

Art and Bri: Focuses on sustainable living, gardening, and homesteading tips, along with practical DIY projects.

Beekeeping with Ted: Features informative videos on various aspects of beekeeping, including equipment reviews and troubleshooting common issues.

Bigger Than The Three of Us: Focuses on gardening, DIY projects, and homesteading, with practical advice for beginners.

Deep Roots Homestead: A channel dedicated to homesteading life, including gardening, livestock care, and preserving food through canning.

Farm Alarm: A homesteading vlog that covers various topics, including animal husbandry, gardening, and self-sufficiency tips.

Fermenting for Foodies: Offers practical tips and tutorials for beginners looking to explore fermentation at home.

Green Dream Project: Offers tips on organic gardening, permaculture, and sustainable living, showcasing the journey of a young couple building their homestead.

Homestead and Chill: Features easy-to-follow recipes and tips for drying foods at home.

Homesteading Family: Provides in-depth information on various canning methods, recipes, and tips.

Homesteady: Features practical tips on homesteading, including animal care, gardening, and food preservation.

Hoss Tools: Provides expert advice on gardening techniques, tools, and growing strategies, particularly for vegetable gardening.

Justin Rhodes: Covers all things homesteading, from permaculture to raising animals to practical gardening advice.

Kaufman's Bees: Offers practical tutorials on beekeeping, hive management, and honey harvesting.

Kendra's Kitchen: Offers tutorials on canning and food preservation with clear, easy-to-follow instructions.

Living Off Grid: Shares the experiences of living off-grid, including renewable energy solutions, gardening, and survival skills.

Living Traditions Homestead: A couple shares their experiences raising animals, growing food, and becoming more self-sufficient.

Off Grid with Doug and Stacy: Chronicles their off-grid lifestyle, sharing tips on homesteading, gardening, and living sustainably.

Red Poppy Ranch: A family-run channel that shares insights on raising livestock, gardening, and DIY homesteading projects.

Roots and Refuge Farm: A down-to-earth channel offering tips on gardening, homesteading, and building self-reliance.

Sow the Land: Chronicles the journey of a family as they build their homestead, sharing tips on animal husbandry, gardening, and sustainable living.

The Grow Network: Offers a wide range of videos on homesteading topics, including gardening, foraging, and self-sufficiency practices.

The Kitchen Garden: Provides visual tutorials on various drying techniques, including using an oven and dehydrator.

The New Kitchen: Features a variety of fermentation recipes and processes, making it easy to follow along.

The Permaculture Orchard: Focuses on permaculture principles and practices, particularly related to fruit trees and sustainable gardening.

The Prepared Homestead: Offers advice on self-sufficiency, gardening, and preparedness, with a focus on sustainable living.

Online Courses

Coursera: Offers courses on resilience-building and problem-solving skills, which can enhance your ability to cope with challenges.

EdX: This platform offer courses on sustainable agriculture, permaculture design, and environmental science from various universities.

Skillshare: Offers courses on various homesteading skills, including gardening, animal husbandry, and DIY projects. These classes can stimulate curiosity and creativity.

The Great Courses: Offers courses on homesteading topics such as gardening, food preservation, and self-sufficiency skills.

The Online Homesteading Academy: This platform provides courses specifically focused on various homesteading skills, from raising chickens to preserving food.

Udemy: Explore courses on specific homesteading skills like gardening, soap making, and sustainable living practices.

ABOUT THE AUTHOR

Audra Haldane is a lifestyle writer and passionate advocate for self-sufficiency whose expertise spans from the sprawling landscapes of Appalachia to the misty hills of Scotland, where she operates her own farms.

Known for her accessible, hands-on approach to homesteading, Audra's books—including *The Homestead Blueprint: Introductory Self-Sufficiency With Homesteading*, *Homesteading Mastery: Advanced Skills and Techniques*, and *Sustainable Homestead Design and Permaculture Principles*—have become essential reads for those looking to cultivate independence, resilience, and sustainability.

When she's not tending to her farms or sharing practical homesteading wisdom, Audra indulges her love for travel and reading, constantly seeking inspiration from new places and stories.

With her guidance, readers from urban high-rises to remote countryside homes discover the joys of sustainable living, making her one of the most beloved voices in the modern homesteading movement.

INDEX

A

B

C

D

T

U

V

W

Y

Z

Index

www.ingramcontent.com/pod-product-compliance
Lightning Source LLC
LaVergne TN
LVHW020234210926
848481LV00016B/1968
9798230927945